3 Receptors

Edited by
P. Cuatrecasas
Wellcome Research Laboratory,
Research Triangle Park, North Carolina

and

M. F. Greaves
ICRF Tumour Immunology Unit,
University College London

and Recognition
Series A

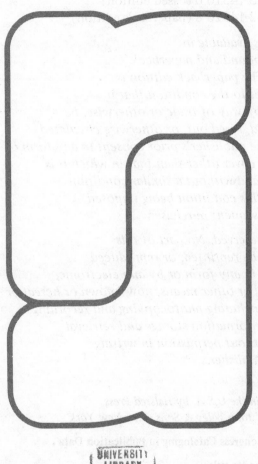

LONDON
CHAPMAN AND HALL

A Halsted Press Book
John Wiley & Sons, Inc., New York

First published 1977
by Chapman and Hall Ltd
11 New Fetter Lane, London EC4P 4EE

© 1977 Chapman and Hall Ltd

Typeset by C. Josée Utteridge of Red Lion Setters
and printed in Great Britain at the University Printing House, Cambridge

ISBN 0 412 14310 0 (cased edition)
ISBN 0 412 14320 8 (paperback edition)

Distributed in the U.S.A. by Halsted Press,
a Division of John Wiley & Sons, Inc., New York

Library of Congress Cataloging in Publication Data

Main entry under title:

Receptors and recognition, (series A)

   Includes bibliographical references.
   1. Cellular recognition. 2. Cell receptors.
I. Cuatrecasas, P. II. Greaves, Melvyn F.
QR182.R4   574.8'761   75–44163
ISBN 0-470-99146-1

Contents

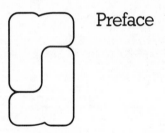

Preface

Lindstrom provides a lucid description of a fascinating new area of membrane research — antibodies to receptors. It is clear that such antibodies offer a valuable tool for receptor studies and also provide some important insights into certain human auto-immune conditions which may involve an immunological modification of receptor function. Injection into rabbits of either purified acetylcholine receptors (from electric eels) or antibodies to this receptor induces a neuro-muscular disturbance which is strikingly similar to the human disease myasthenia gravis. It seems likely that myasthenia in man also involves anti-acetylcholine receptor antibodies since they are detectable in the serum of almost all patients with the disease. Why these patients should react against their own receptor proteins is unknown but is presumably related to some form of immunologic dis-regulation reflected in the concurrent presence of other auto-immune diseases in the same patients.

These fascinating studies prompt enquiry into whether antibodies to other physiological receptors might be of importance in human disease. Lindstrom discusses briefly the suggestive evidence for anti-TSH receptors in thyroid disease and the preliminary evidence for anti-insulin receptor antibodies in a few patients with diabetes. Another possible anti-receptor system, of particular interest to immunologists, involves anti-immunoglobulin idiotype antibodies; these are specific for the antigen receptors on the surface of individual clones of B or T lymphocytes as well as the secreted antibody product of B cells and their plasma cell descendents. Such antibodies can easily be made experimentally and N.K. Jerne has postulated that they might occur 'normally' and serve a physiological feedback role in immune responses.

Precisely how anti-receptor antibodies exert their effect is unclear although studies *in vitro* clearly demonstrate that some may directly competitively block the binding of the physiological ligand. As Lindstrom discusses in the case of the anti-acetylcholine receptor, antibodies to different parts of the receptor structure may have differing effects. Certainly the antibodies themselves are heterogeneous in specificity and in patients with thyroid disease it is recognised that antibodies to both receptor and adjacent but non-receptor membrane structures may exist. The effects of such antibodies on target tissue is clearly not predictably *a priori*; some block normal function (as in the case of myasthenia) by obscuring or blocking the ligand-combining site or by receptor removal (modulation), whereas others may actually stimulate and thereby mimic a normal hormone function (e.g. long-acting thyroid stimulator or LATS antibody).

Marjorie Crandall discusses sexual interactions in the microbial world. Although the genetics and general biology of microbial mating has for many years provided intriguing evidence for the existence of a diversity of highly specific systems for ensuring appropriate complementarity of mates it is only very recently that this subject has been studied at a molecular level. Despite the great diversity of microbial arrangements for several interactions described by Dr Crandall , readers of Lewis' review Incompatibility in Flowering Plants in Vol. 2 of this series will recognise the basic similarities to the systems developed and diversified by higher plants.

There is also perhaps more than a superficial resemblance to the histocompatibility systems of higher vertebrate animals (see review by Greaves in Vol. 1). All the various genetic and molecular arrangements for mate attraction and contact appear to be directed towards at least one principal goal – ensuring heterozygosity of offspring; this demand clearly relates to the value of genetic diversity (i.e. hybrid 'vigour') and seems to be achieved in fungi and plants at least by selection for *non-identity*. McFarlane Burnet (see Vol. 1 in this series, p. 35), has pointed out that similar reproductive interactions almost certainly occur in invertebrates and they could have an evolutionary link with immuno-logical recognition since both reproduction and immunological inter-action require a self-non-self discrimination.

Heinz Furthmayr has provided an authoritative up to date report on the proteins of the red cell surface membrane. The red cell glycoprotein, termed 'glycophorin' by Furthmayr's colleague Marchesi, is one of the few membrane proteins that have been extensively studied and whose structure and orientation in the membrane are reasonably well-defined. The study of glycophorin and other red cell proteins therefore provides something of a precedent and a model for other systems. Red cells are of course very different from most cells and their membrane structure is clearly different, although perhaps not in any fundamental sense, from that of, say, lymphocytes*. The red cell may or may not provide us with a model for understanding receptor function; however it is already clear that studies on this cell are providing a major contribution to our understanding of membrane structure. Dr Furthmayr's article, although

---

* Bretscher, M.S. and Raff, M.C. (1975) Mammalian plasma membranes, *Nature*, **258**, 43.

providing a very detailed analysis of red cell membrane proteins embodies a succinct and critical synthesis of the general relevance of the cellular system under study. The review is written in an admirably clear form which less chemically orientated biologists should have no difficulty in comprehending.

Dr Silverman who has carried out extensive research on membrane transport of sugars in the dog kidney discusses the molecular details of both this system and others which transfer small molecules across membranes. This type of function is a ubiquitous cell surface activity and it is fascinating to see how the details of membrane transport mechanisms have evolved rapidly in recent years into the general concepts of membrane structure. There is clearly a very considerable diversity of transport proteins for different transportable molecules and marked differences from tissue to tissue and between one species and another. Perhaps the most interesting aspect of this subject is the relationship between the specificity of the transport protein and the specificity of the intracellular metabolic activities which the transported molecule engages in. This raises some interesting possibilities with respect to evolutionary relationships between carriers and enzyme. It is intriguing to note that studies of microbial chemoreception, particularly in bacteria, have recently focussed upon essentially similar questions (see 'Receptors and Recognition' Series B, *Microbial Interactions,* edited by J. Reissig).

January 1977   P. Cuatrecasas
                                                              M.F. Greaves

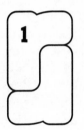

Antibodies to Receptors for Acetylcholine and other Hormones

JON LINDSTROM
The Salk Institute for Biological Studies,
San Diego, California 92112

Abbreviations

AChR — acetylcholine receptor
MG — myasthenia gravis
EAMG — experimental autoimmune myasthenia gravis
αBGT — α-bungarotoxin
anti-AChR — antibody to AChR
mepp — miniature endplate potential
epp — endplate potential
Fab — combining site fragment of IgG
IgG — immunoglobulin G
LATS — long-acting thyroid stimulator
TSH — thyroid-stimulating hormone

Acknowledgements

Electronmicrographs were generously provided by Andrew Engel. Valuable discussions with colleagues have contributed to some of the ideas discussed in this chapter. These colleagues include: A. Engel, E. Lambert, S. Heinemann, M. Seybold and V. Lennon. My research has been supported by the National Institutes of Health (NS — 11323) and the Muscular Dystrophy Associations of America. I thank S. Fuchs, A. Karlin and B. Fulpius for sharing their manuscripts prior to publication.

Receptors and Recognition, Series A, Volume 3
Edited by P. Cuatrecasas and M.F. Greaves
Published in 1977 by Chapman and Hall, 11 Fetter Lane, London EC4P 4EE
© 1977, Chapman and Hall

1.1 INTRODUCTION

Signaling between cells is frequently accomplished through molecules released by one cell which bind to specific receptors in the surface membrane of another cell. Biochemical studies of these receptors are often difficult because the receptors are usually present in minute amounts, and because the function they perform is often difficult to define after they have been removed from the membrane.

Antibodies to receptors are an especially useful tool for studying the molecular biology of receptors. Large amounts of antibodies can be generated by immunization with small amounts of purified receptor protein. These antibodies can bind to both defined structural components on the purified receptor and to these components on receptor in intact tissue. If binding of antibody to receptor alters receptor function, as is frequently the case, the antibody can be used as a probe for assessing the functional role in intact tissue of structures defined in the purified protein. Autoantibodies to receptors are involved in several disease states. They provide an *in vivo* probe for studying the molecular and cell biology of receptors. Because receptor structure is fairly conservative despite substantial evolutionary distance between species, there is some degree of cross-reaction between antibodies to receptors from one species and the analogous receptors from another species. Thus antibodies to receptors can also be used to compare receptors in one species with those in another.

This chapter will focus primarily on studies of the nicotinic acetylcholine receptor (AChR) because this is the best-studied receptor pharmacologically, electrophysiologically, biochemically, and immunologically. The AChR has been purified [1−11] and its molecular substructure is under study. Antibodies formed in rabbits against AChR purified from electric eels block activity of AChR on eel electric organ cell *in vitro* [12]. The antibodies also effect the function of AChR in skeletal muscle of the rabbits causing weakness and death [13]. We have termed the disease produced in animals by immunization with AChR 'experimental autoimmune myasthenia gravis' (EAMG) [14] because of its

3

similarity to the human disease myasthenia gravis (MG). Now antibodies to AChR have been found in patients with MG [15—22]. In both the animal model and MG the effects of autoantibodies are much more complex than simply preventing binding of acetylcholine to AChR. In addition to inhibition of AChR activity [23—26], changes in receptor metabolism [27—29] and membrane structure occur [30—33]. These antibody-induced changes provide insights into the molecular biology of neuromuscular transmission.

Antibodies to some other receptors will also be reviewed briefly. In some of the examples chosen, autoimmune diseases are involved. The range of antibody effects on these receptors is really quite fascinating. In one case, antibody binding to receptor turns the receptor off; in another case, it turns the receptor on. In one case, antibody prevents ligand binding, but in another, antibody increases ligand binding. In yet another case, antibody binding is completely without effect on function. In every case the antibodies provide critical insights into how the receptor works.

1.2 SOME GENERAL REMARKS ABOUT RECEPTORS

The word 'receptor' as used in this chapter refers to a surface membrane component, usually a protein, which regulates some biological event in response to reversible binding of a relatively small ligand. The 'receptor' is usually taken to comprise both the ligand binding site and the molecular machinery which is directly regulated by ligand binding, for example, an ion channel through the membrane. The primary event which ligand binding regulates is often far removed from the most readily observable events controlled by the receptor — a long chain of events intervenes between binding of acetylcholine to AChRs and contraction of muscle or between binding of insulin and increased glycogen synthesis, increased glucose metabolism, and other effects. It is often difficult to define what the primary event regulated by ligand binding is. For example, does insulin inhibit an adenylcyclase or activate a guanylcyclase, or does it have some other primary effect which in turn alters cyclase activity? Purification of functional receptors is the most direct means to solving that sort of problem. Even when the primary event is reasonably certain, substantial problems remain. For example, the nicotinic AChR in skeletal muscle appears to include a cation-specific channel. Solubilization of AChR from the membrane prevents its primary activity from being measured by the normal electrophysiological method. Therefore, it must be purified first and only then tested by reconstituion into a membrane to determine whether the channel has been recovered intact.

The signal transmitted by a hormone or neurotransmitter depends on the nature of the receptor. There can be several different receptors for the same molecule in an individual. For example, the thyroid stimulating hormone (TSH) receptor in thyroid and the TSH receptor in retro-orbital tissue recognize different parts of the TSH molecule and can be activated independently with modified TSH molecules [35]. Acetylcholine is an excitatory neurotransmitter on skeletal muscle where binding to receptor causes a rapid increase in permeability to both Na^+ and K^+ resulting in depolarization of the membrane [137]. On the other hand, acetylcholine is inhibitory to heart muscle where binding to the receptor causes (perhaps as a secondary event to increased guanyl-cyclase activity [37]) increased permeability to K^+ only of relatively slow onset and long duration that hyperpolarizes the membrane [36]. On central nervous system neurons, acetylcholine has been found to be excitatory, but there the effect is to obtain depolarization by decreasing K^+ permeability, and the AChR involved has different drug-binding specificities than the excitatory AChR of skeletal muscle [36]. The excitatory AChR on smooth muscle appears different from those listed before [36]. Clearly, the receptor system discussed at any point must be carefully defined.

Receptors are usually present in very small amounts. They are probably required only in small amounts because in their regulatory function they act as signal amplifiers. For example, the binding of a single acetylcholine molecule to an AChR in skeletal muscle causes a current involving 5×10^4 cations [38]. Although in some cases receptors may be diffusely spread throughout the cell membrane, in other cases they are concentrated into specialized regions. For example, AChR in skeletal muscle are concentrated in parts of the postsynaptic membrane immediately underlying the nerve ending [39]; and, in these parts of the membrane probably comprise the predominant membrane protein. But before innervation or after dennervation AChR are found in lower density, but larger total amount, throughout the surface membrane of the muscle [40, 41].

1.3 THE NICOTINIC ACETYLCHOLINE RECEPTOR OF
SKELETAL MUSCLE AND ELECTRIC ORGANS

1.3.1 Function

Motor neurons usually synapse with skeletal muscle fibers at one point
in the middle of the fiber. The extent and orientation of nerve terminal
branching results in synapses of various configurations depending on the
type of muscle fiber. Ultrastructure of a typical nerve muscle junction
is shown in Fig. 1.1 and 1.2. Acetylcholine is packaged in quanta of less

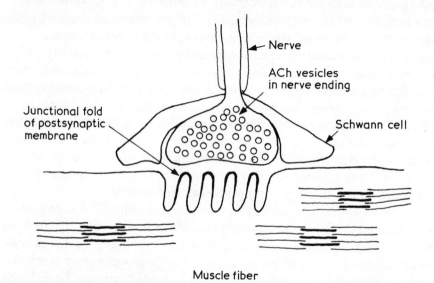

Fig. 1.1 Diagramatic representation of a normal nerve–muscle junction.

than 10^4 molecules [42] contained in vesicles in the nerve ending. The
nerve ending is separated from the postsynaptic membrane of the muscle
by a gap of 600 Å [39]. The postsynaptic membrane is highly folded so
that the tips of the junctional folds where AChR is most concentrated
[39] are advantageously oriented with respect to sites where quanta of
ACh are released from the nerve ending [43]. Acetylcholine esterase
is located along the postsynaptic membrane in association with basement
membrane [44].

An electrical potential of about 70 mV inside negative exists across
the surface membrane of the muscle fiber in the resting state [36, 137].
This results from difference in cation concentrations across the membrane.

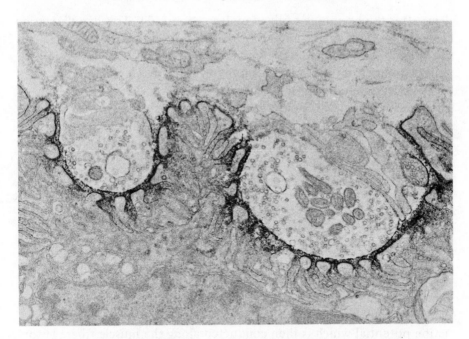

Fig. 1.2 Electronmicrograph of nerve-muscle junction in normal rat forelimb muscle (x 14 500). Dark staining at tips of postjunctional membrane folds is due to the peroxidase reaction product deposited due to peroxidase-αBGT bound to AChR. Some staining is also seen on the presynaptic membrane, probably due to diffusion of the peroxidase reaction product. The tissue was fixed with osmium tetroxide but not stained to increase contrast of the stained portions. This photograph and all others in this chapter were taken by Dr. Andrew G. Engel of the Mayo Clinic. (Reproduced from *Neurology* 33 by permission.)

This membrane is somewhat less impermeable to K^+ than Na^+, and Na^+ is pumped out of the cytoplasm by an ATP-driven enzyme in the membrane, the sodium pump. The net result is that internal K^+ concentration is higher than extracellular, and internal Na^+ is lower than extracellular. The resulting membrane potential is measured with intracellular microelectrodes. Changes in the relative conductance of the membrane to K^+ and Na^+ causes ionic current flow across the membrane and change in the electrical potential.

Quanta of acetylcholine are spontaneously released from the nerve ending at the rate of 1 per second [45]. The acetylcholine released causes depolarization (a reduction in the membrane potential) of the postsynaptic membrane by a maximum of about 1 mV in less than

1 ms after the beginning of the depolarization [42]. This depolarization is termed a spontaneous miniature endplate potential (mepp). Acetyl-choline-induced depolarizations are quickly terminated by diffusion of acetylcholine from the synapse and hydrolysis of acetylcholine by acetylcholinesterase [45]. On the average, an acetylcholine molecule combines with an AChR only once, and the density of AChR in the membrane effected by the quantum is so large that they are not satur-ated by the acetylcholine in a quantum [42]. When an action potential conducted along the motor neuron axon invades the nerve ending and depolarizes it, a number of quanta are released simultaneously resulting in an evoked endplate potential (epp) [45]. In human intercostal muscle about 40 quanta are released from the ending, whereas in rat forelimb muscle, for example, 200 quanta are released [26]. The acetylcholine from these quanta produced a much larger depolarization of the post-synaptic membrane. The depolarization spreads electrotonically along the membrane. If the number of AChRs activated is sufficiently large, the epp, by the time it spreads out of the endplate into the adjacent electrically excitable membrane, will still be large enough to trigger an action potential which is then conducted along the muscle fiber. Through the processes of excitation-contraction coupling, the action potential causes release of Ca^{2+} into the sarcoplasm which in turn activates the actomyosin interaction that actually produces muscle contraction. Clearly, the binding of acetylcholine to AChRs is far removed from the final effects of its action.

The interaction of single acetylcholine molecules with AChRs can be studied electrophysiologically. Bath application of acetylcholine to a muscle causes an increase in the electrical noise recorded by an intra-cellular microelectrode due to activation of AChRs. Complex analysis of this acetylcholine noise suggests that binding of acetylcholine to an AChR opens an ion channel permitting a current of 10^{-11} A over a period of about 1 ms (corresponding to the flow of $\sim 5 \times 10^4$ ions) [38]. This channel is thought to be about equally permeable to K^+ and Na^+ [36]. Opening the ACh channel short circuits the membrane potential and permits passive flow of ions along their concentration gradients. Curare is a competitive antagonist of AChR. Curare prevents binding of acetylcholine and stabilizes the AChR in its resting conformation, thereby reducing the number of acetylcholine noise events without altering those that remain [38]. As determined by pharmacological techniques, carbamylcholine has lower affinity for AChR ($K_D = 3 \times 10^{-5}$ M) than does acetylcholine ($K_D = 2 \times 10^{-6}$ M) [46]. The noise produced by carbamylcholine is of shorter duration, as if the channel remained open for the time that ligand was bound [36]. Other agonists differ from ACh in both duration and conductance of elementary events [47], suggesting that AChR is extremely conformationally sensitive. Reduction of a disulfide bond located in or near the acetylcholine binding site changes the action of hexamethonium from antagonist to agonist [48], and reduces the amplitude of acetylcholine noise [49]. Opening and closing of AChR channels is even sensitive to the potential across the membrane [50, 51]. Also, the AChR shows desensitization, meaning that the membrane conductance decreases to normal in the presence of continued high concentrations of ACh. The desensitized conformation of AChR is thought to exhibit high binding affinity for ligands [52]. Relatively low affinity binding of acetylcholine is expected if the lifetime of acetylcholine—AChR complexes is to account for the brief conductance events observed electrophysiologically [53].

Electric organ tissue from the fresh water electric eel, *Electrophorus electricus*, or the marine ray, *Torpedo*, is apparently an evolutionary derivative of nerve muscle junctions. The large multinucleate cells in these electric organs have no muscle-like contractile proteins. Their surface is heavily innervated with of the order of 10^5 synapses per cell [54]. The cells are arrayed in stacks like cells of a storage battery, and electrical discharges are achieved by synchronous discharge of the membrane potential of these cells. *Torpedo* has a higher concentration of AChR because all current passes through AChRs when these cells are stimulated, whereas *Electrophorus* cells are also electrically excitable and so current flows through both AChRs and action potential channels [55]. AChRs in *Electrophorus* appear pharmacologically and electro-physiologically quite similar to the AChRs in skeletal muscle [46]. AChR in *Torpedo* is less well characterized in these respects [46].

1.3.2 Structure

Electric organ tissues has proved a rich source for purification of AChR. A muscle fiber has a single synapse containing $\sim 4 \times 10^7$ AChR [56], whereas an eel electric organ cell contains $\sim 2 \times 10^{11}$ AChR [57]. Even so, the yield of AChR from eel is small (typically 3—5 mg protein kg electric [11]) and substantial purification is required. *Torpedo* contains larger amounts of AChR at greater concentration, so the yield is much greater (40 mg protein/kg organ [58]). Biochemical studies of AChR depend on the use of snake venom toxins which bind to the acetylcholine binding site of AChR with high affinity and great specificity [59]. The most commonly used toxins are α-bungartoxin from the krait, *Bungarus multicinctus,* toxin 3 from the cobra, *Naja naja siamensis,* and toxin from the spitting cobra, *Naja nigricollis.* All are basic proteins. α-Bungarotoxin (αBGT) has 74 amino acid residues cross-linked by 5 disulfide bonds [59]. It has the highest proportion of non-polar amino acids and binds the least reversibly [59]. Toxin 3 from *N. naja siamensis* is similar but has 71 residues, whereas the toxin from *N. nigricollis* is typical of another set of cobra toxins and has only 61 residues [59]. Labeled toxin can used to localize AChRs in tissue. Specificity of toxin binding is demonstrated by inhibition of binding with AChR ligands like curare and by showing that toxin binding, like AChR, is restricted to endplates in normal muscle, but spreads over the fiber surface after denervation [60, 61]. Using αBGT labeled with [125]I [39] or peroxidase [33], fine scale localization of AChR in the postsynaptic membrane has been achieved. AChR is most concentrated at the terminal expansions of the folds in the postsynaptic membrane [39].

AChR can be solubilized from membranes in a form which retains the ability to bind toxin and cholinergic ligands using mild detergents like Triton X-100 [1–11]. These detergents displace the membrane lipids normally surrounding AChR. Unfortunately, they also alter AChR conformation. Even low concentrations of Triton X-100 inhibit AChR activity in cells [62]. Binding affinities for cholinergic ligands change from those in intact cells when the membrane is fragmented, and change still further during solubilization and purification [53]. Activity of the AChR ion channel could not be assayed after solubilization, so AChR was identified and purified on the basis of its ligand-binding properties.

AChR has been purified by affinity chromatography from both *Electrophorus electricus* [1–3, 8–10], several species of *Torpedo* [4–7, 63] and mammalian muscle [11, 27]. One approach to affinity chromatography has been to use neurotoxin from *Naja naja siamensis* conjugated to agarose beads as an affinity absorbent from which the specifically bound AChR are specifically eluted by a cholinergic ligand [1, 6, 9, 10]. The other approach has been to use a synthetic cholinergic analogue on the agarose beads to adsorb AChR and then salt gradients or a cholinergic ligand to elute the bound AChR [2–5, 8, 11, 63].

Purified AChR is a protein [1–11] containing some carbohydrate [64, 65, 137], but no associated lipid [65]. AChR from eel [1, 2], like that from muscle [61], sediments at 9.5 S. AChR from *Torpedo* is composed of particles and dimers of this size [17, 66, 137].

Electronmicroscopy of purified AChR and AChR-rich membrane fragments reveals doughnut-shaped aggregations 3—4 nm in diameter of 5—6 subunits with a negatively staining hole in the center [67, 68]. In AChR-rich portions of the membrane these are densely packed with a center to center distance of 9—10 nm. The number and exact molecular weights of the polypeptide chains comprising AChR are not yet universally agreed upon, but the differences probably reflect the different electrophoretic techniques used, differing degrees of proteolysis of AChR during purification, and species differences rather than differences in the identity of the AChR molecules purified by different laboratories from any given species. Typical results are given. SDS acrylamide gel electrophoresis of purified eel AChR reveals two bands (42 000 and 54 000 daltons apparent molecular weight [1]) or three bands, if the sample is not heated before electrophoresis (40 000, 47 000 and 53 000 daltons [69]). Several methods suggest that the 9 S form of AChR corresponds to a molecular weight of 230—295 000 daltons [1, 70, 71, 137]. Electrophoresis of *Torpedo* AChR reveals 4 bands (39 000, 48 000, 58 000, 64 000 daltons [63]). The highest specific activities reported for binding of toxin or acetylcholine suggest the presence of one acetylcholine binding site per 100 000—150 000 daltons of protein [1—11]. Frequently, preparations of equal purity, but lower specific activity are used. Despite this apparent denaturation frequently accompanying purification, the AChR appears to have more than one binding site for acetylcholine per macromolecule. From affinity labeling experiments the lowest molecular weight chain (39 000—42 000 dalton) of both eel and *Torpedo* AChR is known to contribute part or all of the amino acid residues to the acetylcholine binding site of the AChR [63]. The functional roles of the other polypeptide chains are not known, so it cannot yet be stated with certainty that they are components of the physiologically significant AChR, though they probably are. They may be components of the receptor ionophore [1]. Reconstitution experiments suggest that purified AChR contains a functional ionphore. Acetylcholine-sensitive cation permeability has been reported using artificial membranes and purified AChR preparations [72—74, 139]. However, reconstitution studies are at a very early stage, and the results obtained thus far have been variable and not altogether typical of the properties of native AChR.

A diagramatic representation of an AChR is shown in Fig. 1.3. Structural details of this model are not to be taken too seriously. It is merely intended to suggest that the AChR is an integral membrane

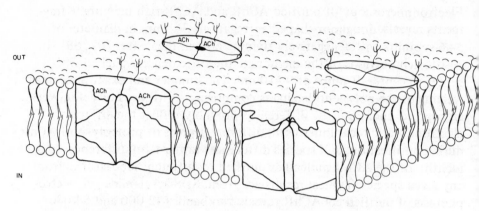

Fig. 1.3 A diagramatic representation of the AChR depicting it as an integral glycoprotein in a postsynaptic membrane composed of a phospholipid bilayer. It is represented as having two binding sites for acetylcholine and one transmembrane ionophore channel whose opening is regulated by a conformational change of the protein. The number of sites and ionphores per AChR macromolecule is not, in fact, known. The AChR is shown to consist of several dissimilar polypeptide chains.

glycoprotein composed of several polypeptide chains with a cation-selective channel that traverses the membrane. The AChR is represented as having more than one ACh-binding site, and these sites are indicated to regulate channel opening by conformation change. The stoichiometry of binding sites to channels or their structural relationship is not known.

1.3.3 Synthesis and destruction

Synthesis and turnover of muscle AChR have been studied. These studies are important to understanding the complex effects of the immune response to AChR on muscle AChR which will be discussed in sections 1.5 and 1.6. Examination of AChR synthesis and turnover, like all AChR studies, are hampered by the small amounts of AChR in muscle, but rather elegant approaches have allowed significant progress. AChR synthesis in cultured myotubes from fetal calf muscle was followed by incorporation of ^{35}S-methionine into the AChR which was then affinity-purified on toxin agarose [75]. AChR synthesis in cultured chick myotubes was followed using deuterium-labeled amino acids to shift the density of ^{125}I-toxin labeled AChR sedimented on sucrose gradients [76].

Denervation of muscle causes synthesis of AChR outside the nerve-muscle junction. Activity is important in regulating synthesis of extra-junctional AChR, and blockage of transmission with curare for 3 days caused a large increase in extrajunctional AChR [77]. Incorporation of ^{35}S-methionine into extrajunctional AChR in organ-cultured denervated rat diaphragms was followed by purification of the AChR on toxin agarose [78]. Little ^{35}S-methionine was incorporated at junctions where the nerve was retained, suggesting a low rate of synthesis of AChR at the junction [78]. Using ^{125}I-αBGT as a functionally irreversible label of AChR, turnover of extrajunctional AChR was found to be much more rapid than turnover of junctional AChR [79]. The half time for loss of extrajunctional AChR has been reported as 8 [79] − 24 [80] hours, whereas with junctional AChR half of the toxin was lost in 7.5 days [81]. It is thought that loss of AChR from the membrane proceeds through internalization and proteolysis [80]. *In vivo* labeling of AChR with αBGT and recovery does not result in any morphological change in the ultrastructure of the postsynaptic membrane, nor any change in the distribution of AChR which can be labeled with ^{125}I-αBGT [82]. Inhibitors of RNA and protein synthesis as well as of energy production can inhibit turnover of AChR [79, 82]. Synthesis and degradation of AChR do not appear to be coupled [80]. At least in myotubes which are rapidly synthesizing AChR, there is an internal pool of AChR equal to about 10% of the surface AChR which can continue to provide AChR to the surface after inhibition of protein synthesis [80].

1.4 ANTIBODIES TO PURIFIED AChR

Immunization of rabbits with AChR purified from eel results in the formation of antibodies. These antibodies were used to demonstrate the purity of the AChR preparation. Immunoprecipitation of AChR by the Ochterlony technique [12, 13] or immuno-electrophoresis [83] gave a single precipitin line. More importantly, these antibodies blocked the depolarizing response of eel electric organ cells to carbamylcholine [12, 15, 84]. This showed that the purified material contained at least part of the physiologically significant AChR.

A radioimmunoassay was developed using ^{125}I-toxin labeled AChR as antigen [12]. After incubation of ^{125}I-toxin labeled AChR with serum, any ^{125}I-toxin-AChR-antibody complexes formed were precipitated by goat anti-rabbit immunoglobulin. Then ^{125}I-toxin in the precipitate was measured. Because the antibody precipitated [125] I-toxin labeled AChR, it was obvious that antibodies directed at determinants other than the acetylcholine binding site were present. This was important because these antibodies might be used as probes for portions of the AChR molecule other than the ACh binding site. Although toxins and a plethora of acetylcholine analogues were available for binding to the acetylcholine binding site, only local anesthetics [53], atropine [85] and histrionico-toxin [86] were available as even potential probes for other parts of the receptor molecule. The acetylcholine binding site of the AChR is inter-esting because it is the trigger, but the ionophore is the barrel of the gun.

In order to detect antibodies which might be inhibited from binding by [125] I-toxin, AChR labeled with [3] H acetic anhydride was prepared [12]. This antigen was not distinguished from unacetylated AChR by the antibody and gave precipitin curves like those obtained with [125] I-toxin labeled AChR as antigen [12]. However, a small inhibition of precipita-tion of [3] HAc-AChR by rabbit anti-AChR was obtained in the presence of excess unlabeled toxin [12]. This suggested that a small amount of the antibody might be directed at or near the acetylcholine binding site of the AChR [12]. This antibody could account for the blockage of AChR activity on electroplax by antibody. However, goat anti-AChR showed no anti-site antibody by this assay [17], and yet was quite capable of blocking AChR activity [84]. Lack of antibodies to the acetylcholine binding site is interesting because it is a determinant exposed on the surface of the AChR molecule which should be readily accessible to immune recognition. Possible explanations are that the site is inherently not antigenic or that it is so nearly identical to the site in the AChRs of the goat that the goat was tolerant to that determinant and unable to form antibodies. The first possibility is more likely since, as will be described later, immunization with eel AChR does cause production of autoantibodies to other determinants on AChR. Under conditions where 75% of AChR on the cells could be shown to be labeled with goat antibody, binding of [125] I-toxin to the cells was inhibited by 15% [84]. This is consistent with the idea that antibodies to AChR can inhibit AChR activity without necessarily inhibiting ACh binding, perhaps by altering the ionophore or regulation of the ionophore by the acetylcholine-binding subunit.

In the future, studies with antibodies raised against each of the poly-peptide chains composing AChR should enable the functional role of these chains to be determined. The low molecular weight chain is involved in binding acetylcholine [69], but whether the other two chains comprise other parts of the AChR, like the ionophore, is unknown. Antibodies raised against purified eel AChR react nearly equally with each of the polypeptide chains in the macromolecule [87]. If antibodies raised against each purified chain retained the ability to bind to native AChR, and if they altered AChR activity in electroplax when they bound, then this would prove that these chains were components of AChR. In addition, the way in which antibody binding modified AChR activity would help reveal the functional role of each chain.

Antibodies raised against AChR from eel cross-reacted to a small but measurable extent (3%) with AChR from *Torpedo* [17]. This cross-reaction occurred at determinants other than the binding site for acetylcholine [17]. When using denatured polypeptide chains from *Torpedo* as antigen, cross-reaction was, however, confined to the 42 000 dalton chain [66] which is known to include the acetylcholine-binding site. Antibodies to eel AChR also cross-react *in vitro* [23, 24] and *in vivo* [23, 28] with AChR on skeletal muscle, with the result that neuro-transmission is impaired. This will be dealt with in detail in section 1.5.

Cross-reaction of rabbit anti-AChR with muscle AChR has been used as a tool in studying synthesis of AChR. Antibody precipitation of ^{35}S-methionine labeled AChR purified from cultures of fetal calf myotubes was used to identify the purified material as AChR [75]. Antibody binding did not prevent toxin binding, indicating that cross-reaction occurred at determinants other than the ACh binding site. Antibody precipitation of ^{35}S-methionine labeled extrajunctional AChR synthesized by organ-cultured denervated rat diaphragms could not distinguish extrajunctional from junctional AChR [78]. This indicated substantial similarities between junctional and extrajunctional AChR, despite the small differences in curare affinity and isoelectric point between them [88].

Anti-AChR has been used for electron microscopic localization of AChR on membranes [143]. Antibodies to AChR purified from *Torpedo californica* were conjugated to electron-dense ferritin molecules. These conjugates were specifically bound on the surface of AChR-rich vesicles purified from *Torpedo* electric organ [143]. Antibodies bound to AChR on the vesicles were also localized using ferritin-labeled anti-IgG. Labeling with ferritin anti-AChR may allow finer localization of AChR in membranes than is possible using peroxidase-labeled αBGT [33]. Diffusion of the electron-dense peroxidase reaction product may lead to erroneous interpretation of the site of peroxidase αBGT binding.

On the other hand, the large molecular weight of ferritin-antibody conjugates may prevent adequate penetration for labeling AChR on the postsynaptic membrane, where peroxidase-αBGT has proven very useful. Also, specific binding of ferritin-anti-AChR to purified membrane fragments containing 20% of their protein as AChR is not so critical a test of binding specificity as is labeling AChR in intact tissue.

1.5 EXPERIMENTAL AUTOIMMUNE MYASTHENIA GRAVIS

1.5.1 Introduction

Immunization of mammals (rabbits [13, 24, 58, 66, 89–91], rats [14, 15, 23, 24, 26–28, 31–33], guinea pigs [14], goats [17], mice [139] or monkeys [92]) with AChR purified from fish electric organs, or even with syngeneic AChR [27] induces an autoimmune response to skeletal muscle AChR which impairs neuromuscular transmission, causing weakness and finally death. Because of the close similarities between these animals and humans with myasthenia gravis (MG), this model disease has been termed experimental autoimmune myasthenia gravis (EAMG).

EAMG was first discovered in rabbits immunized with purified eel AChR [13]. Rabbits were immunized with 400 ug of AChR in complete Freund's adjuvant and two weeks later given a second identical injection. In about two weeks after the second injection the rabbits became weakened, showed a hunched posture, and died within days of respiratory failure. Electromyography demonstrated that there was a defect in neuromuscular transmission. In this technique supramaximal stimulation of nerves in the limbs by surface electrodes causes contraction of muscles which is monitored with needle electrodes in the muscles that record the cumulative action potentials in the fibers near the recording electrode. In normal muscle at low frequencies of stimulation the amplitude of the muscle action potential is constant, meaning that every time the nerves fire the muscles fire. But in the immunized rabbits the muscle response decremented [13], meaning that with successive nerve impulses an increasing number of fibers received a signal insufficient to generate an action potential. Injection of an inhibitor of acetylcholinesterase, which increased the concentration and persistence of acetylcholine in the synaptic cleft, prevented the decrement [13]. Weakness, fatigability, and decrementing electromyogram responses repaired by acetylcholine esterase inhibitors are diagnostic features of MG. Serum antibodies to eel AChR were demonstrated by double diffusion in agar [13] and quantitated by indirect immunoprecipitation using ^{125}I-toxin-labeled eel AChR as antigen [12]. These antibodies blocked the depolarizing response of electric organ cells [12] to AChR agonists.

Studies of EAMG in inbred Lewis rats have proven the most valuable [14, 15, 23, 24, 26–28, 31–33, 93, 94]. EAMG is produced by a single injection of eel AChR (typically 15 ug) in adjuvant. A single injection, as opposed to multiple injections, permits the sequence of events following immunization to be followed. Between 8 and 11 days after injection rats exhibit an acute phase of EAMG [14]. They show signs of weakness and fatigability, a hunched posture, and may die. Their weakness rapidly remits, and they appear essentially normal, until 28 to 30 days when a second, chronic phase of EAMG begins which is progressive and fatal [14]. The acute phase of EAMG involved macrophage invasion of the nerve-muscle junction region [31, 32] which appears to be dependent on labeling of the postsynaptic membrane for destruction by antibody [28]. The chronic phase of EAMG involves the humoral immune response exclusively. In the chronic phase of EAMG, which appears to be the best model for human MG, antibodies appear to interfere with neuromuscular transmission directly by binding to AChR [12, 15, 17, 23] and by inducing decreased AChR content [27, 28, 33] and altered postsynaptic membrane morphology [31–33].

1.5.2 Acute EAMG

Electrophysiological studies of rats with acute EAMG revealed marked impairment of neuromuscular transmission. Decrementing electromyograms were observed [26, 93]. In addition, the amplitude of even the first response was lower than normal [26, 93]. Both of these effects could be produced in normal rats by extensive blockage of AChR activity with toxin [93, 95]. This meant that a substantial portion of the muscle fibers were receiving a signal from the nerve too small to trigger an action potential. Studies with intracellular micro-electrodes also revealed signs of functional denervation at some fibers and impaired transmission at the remainder [26, 32]. No endplate potentials were detected in up to 90% of fibers, although the fibers responded normally to direct stimulation [26]. Also, fibers were observed with resting membrane potentials which were lower than normal [26]. These are results which would be expected in fibers where macrophages had engulfed the postsynaptic

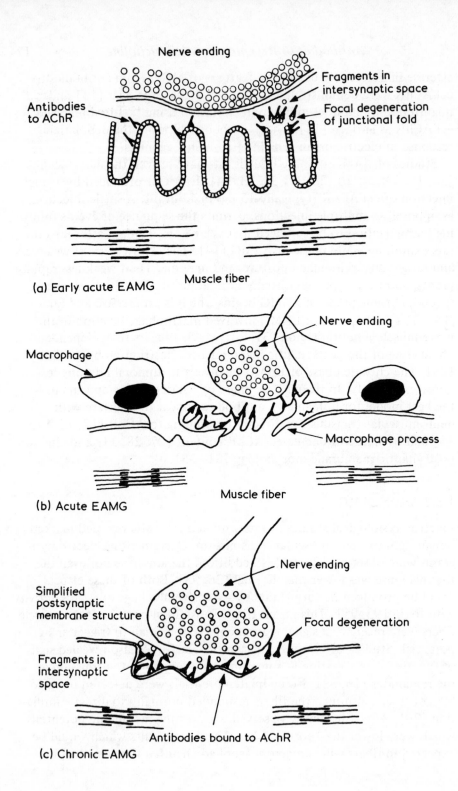

Nerve ending

Fragments in intersynaptic space

Antibodies to AChR

Focal degeneration of junctional fold

Muscle fiber

(a) Early acute EAMG

Nerve ending

Macrophage

Macrophage process

Muscle fiber

(b) Acute EAMG

Nerve ending

Simplified postsynaptic membrane structure

Focal degeneration

Fragments in intersynaptic space

Antibodies bound to AChR

(c) Chronic EAMG

membrane and torn it free from the fiber, thereby preventing transmission and partially shunting the membrane potential.

Evoked endplate potentials detected were of low and variable amplitude [26]. This suggested that effective acetylcholine release was limited to small areas of nerve and postsynaptic membrane which remained in contact despite the cellular invasion. The quantum content of the endplate potentials was in fact reduced to less than 40% of control [26]. The amplitude of mepps which could be recorded was reduced [26]. Forelimb extensor muscle was more severely effected than diaphragm [26]. During recovery from weakness after the acute phase, when neurotransmission had returned to most fibers and the resting potential had returned to normal, the amplitude of mepps remained low and curare sensitivity remained high [26]. However, electromyogram decrements were not observed [26, 93].

The earliest signs of ultrastructural change in the postsynaptic membrane during acute EAMG were focal degeneration at the tips of junctional folds and the appearance of amorphous fragments in the synaptic cleft [31, 32]. These results are consistent with antibody-induced destruction at sites where AChRs are most concentrated. Anti-AChR can fix complement [21, 28], and complement may have mediated some of the membrane damage. Complement factors may also have been involved in causing cellular migration to the effected areas.

Acute EAMG was distinguished by a massive invasion of nerve muscle junctions with mononuclear phagocytic cells and some neutrophil leucocytes [31, 32]. The postsynaptic membrane in many fibers was split away from the fiber, and in some fibers there was segmental

Fig. 1.4 Events in three stages of EAMG are depicted diagramatically.
(a) Indicates the earliest events thought to occur at the nerve-muscle junction during the autoimmune response to AChR. Antibodies are shown to bind to some of the AChR. At one site focal degeneration at the tip of a postsynaptic fold is shown. This might be caused by complement activated by antibodies bound to AChR at this point, and may involve release of chemotactic factors for macrophages. (b) Shows macrophage invasion of the endplate during acute EAMG. Processes of the macrophages are inserted under the postsynaptic membrane during the process of ripping off large areas of this membrane from the muscle fiber. (c) Shows a junction during chronic EAMG. There are no macrophages. The number of postjunctional folds is reduced and simplified. The number of AChR is reduced and many of those remaining have antibodies bound. Focal degeneration continues, and may be responsible for the amorphous fragments in the intersynaptic space.

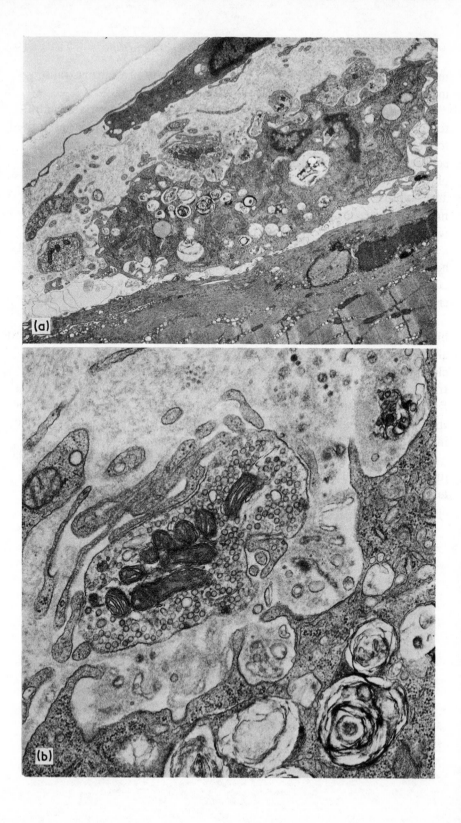

(a)

(b)

necrosis centered on the junctional region [31, 32]. The cellular invasion disappeared as the acute phase remitted. Delayed-type hypersensitivity responses to eel AChR injected intradermally were detected prior to and shortly after the acute phase, but could not be detected in the chronic phase [94]. Whether the cell types responsible for the delayed-type hypersensitivity response are also present at the endplate during acute EAMG is unknown. However, experiments [28] which will be described later indicate that most of the cells involved in acute EAMG are probably non-specific macrophages attacking postsynaptic membrane opsonized with anti-AChR antibodies.

Partial depletion of thymus-derived lymphocytes (T cells) in the rats by injection of antithymocyte serum prevented an obvious acute phase in rats which were then injected with AChR, but the chronic phase did develop [94]. Complete ablation of T cells by thymectomy and X-irradiation, followed by reconstitution with only bone marrow-derived lymphocytes (B cells) prevented development of EAMG when the re-constituted rats were injected with AChR [94]. After reconstitution with both B and T cells, rats developed EAMG when immunized with AChR [94]. These results suggest that formation of anti-AChR antibody requires cooperation between B and T cells, and that interfering with this cooperation by reducing the number of T cells reduces anti-AChR formation shortly after immunization sufficiently to prevent an obvious acute phase.

Low titers of serum antibody to AChR from both eel and rat muscle were detected starting before the onset of acute EAMG [15]. Antibodies recognizing rat muscle AChR at all times were less than 5% of the total [14, 15, 17, 27, 28]. Those antibodies also recognized eel AChR [15]. By 10 days after immunization only one-third of the anti-AChR was IgM, the class usually produced early in immune responses [15]. The remaining two-thirds of the anti-AChR in the serum at 10 days was

Fig. 1.5 Electronmicrographs of a macrophage interposed between the nerve ending and a muscle fiber during the acute phase of EAMG. The entire postsynaptic region has disappeared. (a) (x 4700) The macrophage containing many heterophagic vacuoles separates nerve terminals from the fiber. (b) (x 22 900) At higher power, degraded residues of the postsynaptic membrane folds and remnants of basement membrane are seen interposed between macrophage and nerve terminal. Macrophage processes partially surround this material. Previously phagocytized membrane fragments are seen in vacuoles of the macrophage. (Photo by Dr. A.G. Engel, reproduced from *J. Neuropathol. exp. Neurol.*, **32**, by permission.)

7S IgG. By 35 days all anti-AChR was IgG [15]. Titers of anti-rat AChR increased relatively more slowly after immunization than did titers of anti-eel AChR [15]. Titer of anti-eel AChR reached a maximum before the onset of chronic EAMG, whereas anti-rat AChR had increased to only rather low levels during the acute phase, but increased to much higher levels at the time of onset of chronic EAMG. Antibody was detected bound to AChR in muscle from the start of the acute phase [27].

AChR content of rat muscle decreased transiently during acute EAMG [27]. Much of this decrease no doubt resulted from phagocytic activity. At the time of recovery from the acute phase, content of AChR transiently increased to more than normal levels [27]. This probably resulted not only from repair synthesis as phagocytic cells disappeared from the junction, but also from synthesis of extrajunctional AChR in response to the extensive denervation occurring during the acute phase. Throughout the acute and recovery phases only a small amount of the AChR in muscle had antibodies bound [27]. An increasing fraction of AChR became labeled with antibodies as antibody titers increased and AChR content continued to decrease approaching the chronic phase [27].

1.5.3 Chronic EAMG

Electrophysiological studies of chronic EAMG in rats revealed marked impairment of neuromuscular transmission. Electromyogram amplitude was normal [26, 93] and decrementing electromyograms were observed in only the most severely effected animals [26]. This was attributed to the unusually large number of acetylcholine quanta released by rat nerves [26], which provided a very large safety factor for neuromuscular transmission. Reducing the number of active AChR somewhat with low doses of curare that were without effect in normal rats did cause decrementing electromyogram responses [93]. All fibers showed endplate potentials and resting membrane potential was normal [26]. Small mepps were observed, and evoked endplate potentials could be reduced to levels below the threshold for triggering an action potential by concentrations of curare 1/10 that required in normal rats [26]. These results are probably not due to altered release of acetylcholine, since the number of quanta released was normal [26] and acetylcholine vesicles in the nerve ending and the nerve ending itself were morphologically unaltered [31, 32]. In rats with chronic EAMG, denervated muscle containing extrajunctional AChR was less sensitive to iontophoretic application of acetylcholine [23], showing *in vivo* blockage of AChR. Antisera from rats with chronic EAMG applied to denervated muscle from normal rats caused a reduction in acetylcholine sensitivity showing *in vitro* blockage of AChR by anti-AChR [23].

Rabbits and rats immunized with purified *Torpedo* AChR also showed small mepps and reduced junctional sensitivity to acetylcholine [24]. Rabbit antisera inhibited AChR at frog nerve muscle junctions, causing small mepps of shortened time course [24]. Binding of ^{125}I-αBGT to endplates of the immunized animals was reduced in direct proportion to the serum antibody titer [24].

In rats with chronic EAMG, the structure of the postsynaptic membrane was greatly simplified [31, 32]. Phagocytic invasion was not observed. The number of postsynaptic folds was greatly reduced and total membrane area was reduced. Binding of peroxidase-labeled αBGT to the postsynaptic membrane was greatly reduced [33]. Using peroxidase αBGT, diffuse faint staining was observed as were patches devoid of stain intermittent with patches of more intense staining. This indicated either a reduction in the amount of AChR or blockage of toxin binding (but not necessarily of acetylcholine binding) to AChR by antibodies. Antibody was shown to be bound in the junction using antibody to rat immunoglobulin conjugated to peroxidase [96]. There was evidence of focal degeneration, and fragments of membrane in the cleft and adjacent to it [31, 32] suggested a continuous process of synthesis and destruction of the postsynaptic membrane.

In chronic EAMG high titers of anti-AChR IgG were found [15]. The amount of anti-rat AChR in the serum of a rat with chronic EAMG (typically 5×10^{-10} mol of αBGT binding sites) was far more than the amount of AChR in the rat's muscles (typically 2×10^{-11} mol in a rat with chronic EAMG, which was much less than the 7×10^{-11} mol in a normal rat) [27]. Antibody bound to muscle AChR was demonstrated by extracting the complexes and precipitating them with anti-IgG [27]. The complexes were of a size (> 18 S) suggesting that they consisted of several antibodies and AChRs. These complexes retained the ability to bind αBGT [27], showing that antibodies which bound *in vivo* were not directed at the acetylcholine binding site. More than half, but never 100% of the muscle AChR were bound with antibodies [27].

The AChR content of muscle from rats with chronic EAMG was reduced [27]. Since this could not be attributed to phagocytic activity, this suggested that antibody-induced removal of AChR from the membrane. Antibody-induced removal of surface membrane proteins,

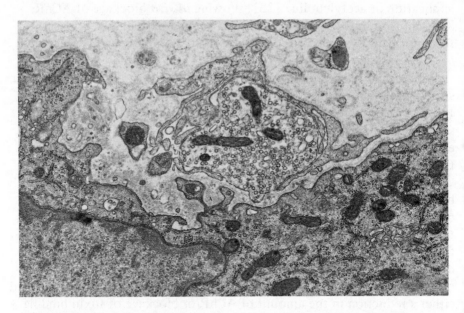

Fig. 1.6 Electronmicrograph of a nerve-muscle junction from forelimb
muscle of a rat in the early stages of chronic EAMG (× 16 900). No macrophages are observed. The postsynaptic membrane is highly simplified. Some
amorphous material is observed between the nerve ending and the few
simplified junctional folds which remain. (Photo by Dr. A.G. Engel,
reproduced from *J. Neuropathol. exp. Neurol.*, **32**, by permission.)

'modulation', is well known in several systems [97]. Observation that
anti-AChR can cause modulation of AChR in cultured myotubes [98]
suggests that this process could occur in EAMG. Binding of antibody to
AChR may cause the complex to turn over by the same rapid process
usually reserved for extrajunctional AChR. Observation of AChR unlabeled by antibody despite the huge excesses present and their ready
access to the junction [27] suggests that these may be newly synthesized
AChR as yet unlabeled. The simplified structure of the postsynaptic
membrane could result from the dynamic balance between increased
synthesis and increased destruction of AChR.

Thymectomy of rats in the chronic phase was not beneficial [94]. This
indicated that, although B-T cooperation was important in induction of
the immune response [94], once B cells were triggered to production of
anti-AChR antibody, T cells in the thymus were no longer important in
the course of EAMG. At even very long times after immunization

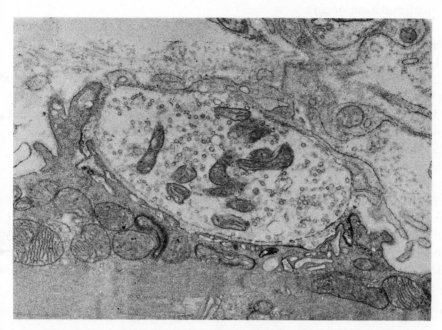

Fig. 1.7 Electronmicrograph of a nerve-muscle junction in forelimb muscle
of a rat later in chronic EAMG (day 38, x 14 900). Darkly staining material
on the postsynaptic membrane is reaction product from peroxidase-αBGT.
Only small areas stain for the presence of AChR. The tissue was fixed with
osmium tetroxide but not stained in order to improve contrast of areas
reacted for peroxidase. (Photo by Dr. A.G. Engel, reproduced from
Neurology, **33**, by permission.)

(> 100 days) titers of antibodies to rat AChR were much lower than the
titers of eel AChR [15]. This suggests that immunization with eel AChR
does not trigger a positive feedback autoimmune response, i.e. rat
muscle AChR effected by cross-reaction with eel AChR did not themselves stimulate further production of antibodies directed exclusively
against rat AChR and not eel. This interpretation would also explain the
failure of thymectomy to suppress chronic EAMG. If no new immuno-
stimulation were occurring requiring B-T cooperation, depletion of
T cells by thymectomy would have no effect. Observation that titers of
both anti-eel AChR and anti-rat AChR decline continuously after about
day 35 [15] suggests that the antibody-producing cells stimulated by the
initial eel AChR immunogen, or the antibodies produced by these cells,
are sufficiently persistent to cause continuing EAMG in even those few
rats which manage to survive for long periods, but that no substantial

autostimulation occurs.

Impaired neuromuscular transmission in chronic EAMG could be caused by any combination of these possibilities:
(1) complete inhibition of AChR activity by bound antibody,
(2) reduction of duration of AChR ionophore opening or its conductance when open due to bound antibody,
(3) reduction in packing density of AChR in the postsynaptic membrane,
(4) reduction in total amount of AChR,
(5) reduction in the amount of acetylcholine reaching AChR due to a larger synaptic cleft or altered juxtaposition of sites of acetylcholine release and AChR concentration.
Probably all these factors are important, particularly (4) and (1) or (2). Because antibodies recognizing an unknown, but probably large, number of determinants on AChR are present, different effects on AChR activity might be expected depending on the specificity and number of antibodies bound to an AChR.

1.5.4 Passive EAMG

It is important to establish the relative contributions of cellular and humoral immunity to the pathogenesis of EAMG. The principal questions are:
(1) are the cells seen in the endplate zone during acute EAMG T lymphocyte killer cells?
(2) is the simplified morphology of the postsynaptic membrane seen in chronic EAMG an obligatory static artifact of the cellular invasion in the acute phase which went before?
The answer is probably 'no' in both cases.

EAMG can be passively transferred with lymphocytes from an immunized animal [94, 99], but this does not define the role of cellular versus humoral immunity since these lymphocytes may well produce anti-AChR that could be the primary effector. AChR are accessible to serum antibodies as shown by their reaction with ^{125}I-toxin coupled to antibodies [100] and by direct measurement of antibody AChR complex in muscle of rats with EAMG [27].

EAMG can be passively transferred to a normal rat by intravenous injection of serum or purified IgG from a rat with chronic EAMG [28]. Transfer is remarkably efficient, and both clinical and biochemical signs can be produced by doses of anti-AChR as low as 1×10^{-11} mol. This amount is only 20% of the AChR content of the recipient and only of the anti-AChR of the donor. This efficiency results from the amplification effect of a massive phagocytic invasion of the junctions that ensues which closely resembles that observed in acute EAMG [28].

Rats injected with 10×10^{-11} mol. of anti-rat AChR showed clinical signs of weakness in 12 h which were identical to those of a rat with EAMG [28]. Weakness was most severe at 36 h, and diminishing, but still present at 4 days. Electromyogram amplitude was decreased, as in acute EAMG [28]. Decrease in electromyogram amplitude followed a time course similar to the clinical signs [28]. Complexes of antibody with muscle AChR were observed [28]. AChR content transiently decreased and then transiently increased to more than normal amounts, during the recovery phase, as in acute EAMG. As in acute EAMG, neurotransmission at forelimb muscles was more severely effected than was transmission at diaphragm muscle [28]. This probably reflected different safety factors for transmission in the two cases, and is reminiscent of the observation in patients with MG that one muscle group is often much more severely affected than another.

These results suggest that, in both passively transferred EAMG and acute EAMG, antibodies bind to AChR on the postsynaptic membrane and tag it for destruction by non-specific macrophages. Binding of cytophilic antibodies to macrophages is probably also important. Antibodies are found on the surface of macrophages in rabbits with EAMG [140]. Binding of antibodies to a few of the AChRs in the postsynaptic membrane results in the destruction of large numbers of AChRs, accounting for the efficiency of passive transfer. This result minimizes the importance of cells specifically sensitized to AChR in acute EAMG, but it raises the question: what terminates the non-specific macrophage invasion in acute EAMG at the time when anti-AChR titers are increasing and might be expected to enhance the invasion?

After passive EAMG, rats do not develop autoimmune EAMG, even if they are given a potent systemic adjuvant [28]. This is an interesting paradox because we know that purified syngeneic AChR is immunogenic in these rats [27], yet when larger amounts of AChR are presented to the receiving end of their immune system in the form of the postsynaptic membrane fragments engulfed by macrophages during passive EAMG, immunity is not triggered. Apparently AChR in its native membrane is not sufficiently altered to break tolerance. Denaturation during solubilization from the membrane, purification and emulsification in adjuvant seems to be required to make syngeneic AChR appear foreign. The transfer of AChR from its native membrane environment to the oil-water emulsion of Freund's adjuvant may be particularly important in this regard.

1.5.5 Immunosuppression of EAMG

Studies of EAMG as a model for developing immunosuppressive therapies
for MG are at a very early stage. Treatment of chronic EAMG rats with
cyclophosphamide prevented exacerbation of their clinical state or increase in anti-AChR titer due to a booster injection of AChR, but did
not reduce their ongoing clinical signs or their initial antibody titer [141]
Rabbits treated with gradually increasing doses of hydrocortisone from
the time of injection with AChR were protected from developing EAMG
[142]. Azothioprine given from the time of AChR injection also prevented development of EAMG, but after termination of azothioprine,
EAMG was readily induced by immunization with AChR [142].

Perhaps chronic EAMG may be a more relevant model for studying
immunosuppressive agents than in preventing development of EAMG in
previously naive animals. Presumably humans with MG are in a constant
state of low level stimulation by endogenous AChR, whereas animals
with EAMG receive only occasional stimulation from depots of exogenous AChR in adjuvant. This may become a troublesome distinction in
developing immunosuppressive therapies for MG using EAMG as a model,
but on the other hand it permits discrimination of induction phenomena
and ongoing effector phenomena. Presumably, the ultimate goal of
studies of immunosuppression of EAMG as a model for testing therapies
of MG would be to take advantage of the well-defined antigen to develop
a method for specifically suppressing immunity to it, while leaving the
overall immune response intact.

1.6 MYASTHENIA GRAVIS

1.6.1 Introduction

Myasthenia gravis is a neuromuscular disease characterized by muscle
weakness which increases with exertion and improves with rest. MG is
twice as common in women as men [101]. The muscle groups effected
vary between individuals. Frequently extraocular muscles are most
effected, resulting in ptosis and diplopia, but in many patients obvious
signs of myasthenia gravis are more generalized. The course of the
disease is often varied, involving spontaneous remissions and exacerbation

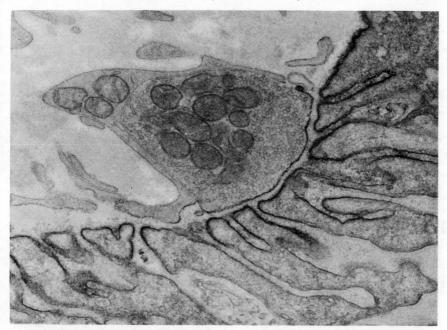

Fig. 1.8 Electronmicrograph of a nerve–muscle junction from normal human intercostal muscle (x 22 300). Dark staining on the terminal expansions of the postsynaptic folds is due to binding of peroxidase-αBGT to AChR localized in these regions. Some staining is also seen on the presynaptic membrane, probably due to diffusion of the peroxidase reaction product. This photograph was taken by Dr. A.G. Engel. (Reproduced from *Neurology, 33*, by permission).

Diagnosis is usually based on subjective evaluation of weakness and improvement after treatment with inhibitors of acetylcholinesterase, and sometimes on objective findings of a decrementing electromyogram response which can be improved by inhibitors of acetylcholinesterase.

Evidence of immunological involvement in MG has been available for a long time. A substantial fraction (5–15%) of MG patients have thymomas and most (70–80%) have thymic hyperplasia [101]. Most patients with thymoma (90%), regardless of whether they have MG, have autoantibodies which bind to structural proteins of skeletal muscle and muscle-like cells in the thymus [102]. Thymectomy may be beneficial for some MG patients [101], and immunosuppressive drugs may also be beneficial [103]. The observation that patients with MG have an unusually high incidence of other autoimmune diseases such as Graves' disease, Hashimoto's disease, and rheumatoid arthritis further suggests the possibility of an autoimmune response in MG [101]. Lymphocyte infiltrations

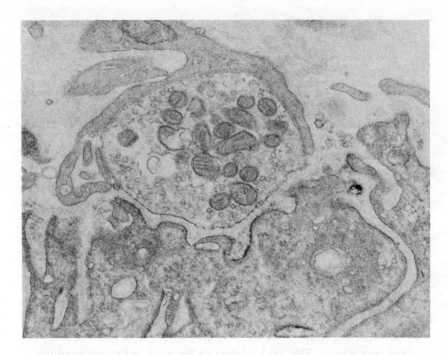

Fig. 1.9 Electronmicrograph of a nerve–muscle junction in intercostal
muscle from a patient with moderately severe generalized MG. Note the
resemblance between the highly simplified postsynaptic membrane shown
here and that in a rat with chronic EAMG shown in Fig. 1.7. Only small
segments of the postsynaptic membrane are stained for AChR with
peroxidase αBGT, showing the decreased AChR content in MG. (Photo
by Dr. A.G. Engel, reproduced from *Neurology*, **33**, by permission.)

are frequently observed in patient's muscles [101]. However, observation
that babies born to myasthenic mothers sometimes exhibit a transient
form of MG, suggests that maternal antibodies might cause the disease
in the baby [101]. Then the disease would remit as the maternal antibodies were cleared from the infant.

John Simpson in 1960 [101] noted this evidence for autoimmune
involvement in MG, and the evidence of extreme curare sensitivity in MG
patients, and proposed that MG might involve autoantibodies to AChR
which competitively inhibited acetylcholine binding. In this section
evidence will be reviewed showing that autoantibodies to AChR are
involved in MG, though not as competitive inhibitors of acetylcholine
binding. There have been several other theories put forth to explain MG.
Goldstein suggested that a factor was released from thymus which

impaired neuromuscular transmission [104], but this factor has not been assayed in MG patients, and there is substantial question [105] about the validity of an animal model developed using autoimmunization with thymus. Small mepps are observed in MG [25, 106], which might suggest a defect in synthesis, packaging, or release of acetylcholine. However, acetylcholine content [107] and quantal content of the endplate potentials [106] are normal in MG, as are the morphology of the nerve ending and the acetylcholine vesicles within it [30]. Furthermore, there is clear evidence of a postsynaptic lesion. Structure of the postsynaptic membrane is greatly simplified [30], it contains reduced amounts of AChR [29, 33, 96, 109] and its sensitivity to acetylcholine is reduced [109]. In the next section evidence will be summarized showing that the lesion inMG doses, in fact result from an autoimmune response to AChR.

1.6.2 Immune response to AChR in MG

Stimulation of thymidine incorporation in peripheral lymphocytes of patients with MG by exposure to purified eel AChR has been reported [110, 111]. In one study 10 of 11 patients showed stimulation beyond the highest control values, with an average stimulation about two fold over background [110]. In another study only 9 of 21 patients showed stimulation [111], these at a level of five fold above background. This indicates a cellular immune response to AChR in the responding patients, but the small number of patients, and the relatively small distinction from control values does not permit a great deal to be made of the stimulation indices observed. Limited reaction between the eel AChR used for stimulation and receptors on cells recognizing human AChR may account for the low and infrequent stimulation. Using this method, no apparent relationship between lymphocyte response to AChR and severity of clinical signs was noted, but patients clinically improving after receiving steroids had a significantly lower response [119].

Antibodies to AChR have been detected in the sera of a high percentage of patients with MG by several methods. In one approach, sera were assayed for their ability to prevent ^{125}I-αBGT binding to AChR solubilized from denervated rat muscle [18]. A maximum of 50% inhibition was detected in 5 of 15 patients in this study [18], but only in 3 of 43 patients in another study [20]. Although using human AChR also permitted detection of some inhibition, this only occurred when anti-AChR was present in large excess over AChR [16]. It was concluded that inhibition of toxin binding to solubilized AChR is not a reliable assay method for anti-AChR [16, 20]. Sera from MG patients are much more effective at blocking αBGT binding to extrajunctional AChRs in frozen sections of denervated human muscle [22, 112]. In this experiment 63 of 92 sera were positive [112]. This method is not quantitative and highly laborious. The tissue was sectioned, labeled with αBGT, fixed, reacted with rabbit anti-αBGT, reacted with goat anti-rabbit IgG labeled with peroxidase, fixed, reacted for peroxidase, dehydrated, embedded, and then observed. Because of the indirect method used to assay αBGT binding, it is uncertain whether the patient's sera prevented binding of αBGT or of the two sandwich layers of antibody used to localize the αBGT. No significant inhibition of ^{125}I-αBGT to sarcolemma fragments from denervated rat muscle was shown by sera from 20 MG patients [20] This suggested that, in the studies of fixed human muscle, binding of αBGT was probably not impaired much by anti-AChR, but instead, binding of anti-αBGT or anti-IgG was impaired. The surface of AChR (2.5×10^5 daltons) exposed on the outside of the membrane might well be sufficient for both anti-AChR (1.5×10^5 daltons) and acetylcholine (182 daltons) or αBGT (7983 daltons) to bind, but not for both anti-AChR, αBGT, and its associated anti-αBGT and anti-IgG.

By far the most effective method for detecting antibodies to AChR in the sera of patients with MG is the indirect immunoprecipitation method used in studies of EAMG, but using human AChR labeled with ^{125}I-αBGT as antigen [15, 16, 17, 113]. This assay detects only antibodies directed at determinants other than the acetylcholine binding site which is labeled by ^{125}I-αBGT. The antibodies detected were of the IgG class [15]. In one study anti-AChR was detected in 62 of 71 sera at an average value 442 fold the average value for normals and patients with other neuromuscular diseases [16]. The cut off for significance was set two standard deviations above the mean for non-myasthenics, and there were no false positive reactions. In another study, anti-AChR was detected in 91% of 144 patients [113]. These results suggest that this method can be used as an objective diagnostic test for MG. The assay is also quantitative, measuring antibody titer in the same units of specifically protectable αBGT binding sites in which AChR is measured.

Several other methods have also been used to detect anti-AChR. Anti-AChR has been detected by indirect immunoprecipitation using ^{125}I-αBGT labeled AChR from denervated rat muscle [19, 20]. AChR from normal rat muscle was not useful as an antigen [15, 18]. In one study using this method 68% of MG sera were found to contain anti-AChR [19], in another 85% were positive [20]. Using rat AChR instead of human, MG patients differed only three-fold [19] from controls instead of 440-fold [16]. Using *Torpedo* AChR as antigen, antibodies were detected in 68% of sera by complement fixation [21]. This provided the useful information that human anti-AChR can fix complement, which indicates a means by which it could induce alteration of the postsynaptic membrane. 65% of MG patient's sera tested prevented binding of rat AChR to Con A sepharose [20]. This means that the anti-AChR bound on or near carbohydrate groups on the AChR. In this study anti-AChR were extracted from MG patients thymuses. Of these extracts 44% prevented binding to Con A sepharose, but 72% contained anti-AChR detectable by immunoprecipitation [20]. This suggests that different antigenic determinants on AChR are recognized by different MG patients [20].

Using the indirect immunoprecipitation assay with human AChR as antigen, a wide range of titers was observed [15, 16, 113]. Titer did not correlate closely with disease intensity in the population studied, though patients with only ocular signs had significantly lower titers [16]. Serial studies of changes of titer in individuals during remission and exacerbation were not done. The average amount of anti-AChR in serum was several fold that required to bind all the AChR present in the patient's body [113]. MG was passively transferred to mice with IgG from MG patients [114]. This shows that, as in EAMG, anti-AChR acts *in vivo* as an effector of the immune response to AChR. Cellular invasion of mouse endplates was not reported. Because interspecies transfer of MG was much less efficient than transfer of EAMG between inbred rats [28] and required large doses and prolonged times, it may be that not only was there limited cross-reaction between anti-human AChR and mouse AChR, but also limited cross-reaction between human anti-AChR and mouse macrophages and complement. Anti-AChR was detected in the sera of babies with neonatal MG at titers less than the mothers [15, 16, 113, 115, 116]. Anti-AChR titers diminished in the neonate during remission, but remained high in the mothers [115, 116]. This provides an example of inter-human passive transfer, illustrating the importance of anti-AChR to the pathology of MG.

Anti-AChR has been detected bound to 36–52% of the AChR in muscle from patients with MG using the same immunoprecipitation method used to demonstrate this in EAMG [29]. Reduced amounts of AChR were also detected [29], as in EAMG [27]. Patients with the least AChR also had the smallest mepps [29]. The early observation that binding of ^{125}I-αBGT to nerve muscle junctions in patients with MG was only 10–30% of normal had suggested loss of AChR [108]. Later evidence that the anti-AChR does not prevent binding of αBGT [20] reinforced this interpretation. Ultrastructural localization of the AChR remaining was achieved using αBGT labeled with peroxidase [33]. The fractional area of the postsynaptic membrane which could be labeled by peroxidase-αBGT was directly proportional to mepp amplitude in these patients. This strongly suggests that, as in EAMG, loss of AChR through an antibody-induced modulation-like phenomenon is the major contributor to impairment of neuromuscular transmission in MG.

1.6.3 Cause of the immune response to AChR

The cause of the autoimmune response to AChR in MG remains unknown. The observation that MG patients have an unusually high frequency of other autoimmune diseases suggests that there may be a predisposing genetic defect in their immune system [101]. Presumably, a specific event triggering autoimmunity to AChR would also be involved. Thymic involvement in MG suggests that the AChR found on the myoid cells in the thymus [15, 117] may play some role in the disease process. But the role might be simply that of passive object of the immune response. AChR in postsynaptic membrane does not seem to be immunogenic since:

(1) passive EAMG does not induce autoimmune EAMG [28],

(2) EAMG does not seem to involve a positive feedback autoimmune response continually stimulated by muscle AChR, and

(3) degenerative diseases of human muscle do not result in anti-AChR formation [15, 16].

Therefore, altered AChR might be a triggering factor in MG. This could occur as a result of infection by a virus which modified cell surface proteins, or it might result from a host-mediated surface modification. Another possibility is that a somatic protein of an infecting virus or bacteria might cross-react with AChR and function analogously to eel AChR in EAMG. Some peculiarities of autoimmune diseases might be interpreted as suggestive of a viral role in modifying host antigens. Autoimmune responses can be arranged in a spectrum ranging from organ-specific (such as MG or Hashimoto's thyroiditis) to non-organ specific (such as systemic lupus erythmetosis, where antinuclear antibodies are found) [120]. Patients with organ-specific diseases have an increased tendency to develop cancer of the effected organ (consider the high incidence of thymoma in MG where the autoimmune response is probably effecting myoid cells) [120, 121]. On the other hand, patients with non-organ-specific disease are more likely to develop generalized lymphoreticular neoplasia [120, 121]. The organ specificity, or lack of it, might arise from the host range specificities of an infecting virus. Neoplasia could arise from occasional viral transformation of the infected cells.

1.7 ANTIBODIES TO OTHER RECEPTORS AND ASSOCIATED DISEASES

1.7.1 TSH receptor

Autoantibodies to the receptors for thyroid stimulating hormone (TSH) are found in up to 85% of patients suffering from Graves' disease (a form of hyperthyroidism) [122]. There are many similarities between the effects of this antibody and anti-AChR, but there is the interesting difference that antibody to the TSH receptor activates rather than inhibits the TSH receptor [122]. When this antibody was first discovered in patients' serum it was termed long-acting thyroid stimulator (LATS).

LATS is an IgG [122]. Both LATS and its monovalent and divalent Fab fragments stimulate thyroid cells through an adenyl cyclase-mediated mechanism as does TSH [120, 121]. TSH is a glycoprotein hormone composed of two dissimilar 13 000–14 000 dalton polypeptide chains [35]. LATS probably binds to the receptor for TSH in or near its TSH binding site, since it can prevent binding of TSH to thyroid membranes [120]. LATS can cross the placenta and cause a transient form of neonatal hyperthyroidism [123] which, like neonatal MG, resolves after a few weeks as the maternal LATS antibody is catabolysed in the infant. Like anti-AChR, LATS shows some interspecies cross-reaction, but not all sera containing LATS that will effect human cells have an effect on mouse cells [122].

IgG autoantibodies to the TSH receptor are also found in patients with malignant exopthalmos (protrusion of the eyeballs), but these antibodies effect only the TSH receptors of retro-orbital tissue and not those in thyroid tissue [124, 125]. The antibodies increase TSH binding to retro-orbital tissue, but have no effect on TSH binding to thyroid plasma membranes [124, 125]. Conversely, LATS antibody does not effect the TSH receptor in retro-orbital tissue. Antibodies from exophthalmos patients bind to TSH receptors only when TSH is bound

[125]. It is unknown whether the antibody binds to TSH and increases the affinity of the complex for the receptor or whether the antibody binds to the receptor itself at another site dependent on a TSH-induced conformational change. These studies were performed using bovine tissue, showing that the human autoantibodies cross-react to some degree with the appropriate receptors in other species.

TSH receptors from thyroid and retro-orbital tissue appear similar in their binding properties for TSH, and both regulate adenylate cyclase activity [126]. The structural difference between these receptors indicated by differential effects of the two types of autoantibodies is also shown by biochemical experiments. Trypsinization of thyroid plasma membranes solubilizes a sialo-glycoprotein (30% carbohydrate 10% sialic acid) of apparent molecular weight 24 000 daltons which retains the binding properties of the intact receptor [124, 127]. Similar treatment of retro-orbital tissue releases a fragment of 75 000 daltons or more [125]. Sialic acid appears important to TSH receptor activity because neuraminidase treatment of the receptor inhibits TSH binding. Because specific gangliosides bind to the β subunit of TSH and prevent its binding to receptor, it has been suggested that the receptor resembles a ganglioside and that the β subunit of TSH binding to the receptor induces a conformational change in the α subunit which can then interact with adenylate cyclase and alter ion transport [128]. It was further proposed that other hormones like leutinizing hormone share a common or similar α subunit for interaction with the cyclase and differ in their β subunits, each of which recognize a different carbohydrate sequence.

Human exophthalmos appears to involve both autoantibodies to retroorbital TSH receptors and defective TSH molecules lacking the C terminal end of the α subunit [125]. An animal model of exophthalmos in guinea pigs has been made using pepsin-treated TSH lacking the C terminal end of the α subunit [125]. The modified TSH stimulates retro-orbital tissues to produce glycosaminoglycans resulting in exophthalmos. Although the modified TSH binds weakly to TSH receptors in thyroid, it does not activate them [35]. There is no autoimmune component in the model disease.

1.7.2 The insulin receptor

Insulin is a protein hormone which binds to receptors in the plasma membrane of the cells it effects. Its many biological effects on the metabolism of sensitive cells appear to be subsequent to a primary effect

on guanylate cyclase [129]. The insulin receptor has been solubilized
in detergent and substantially purified by affinity chromatography [130],
but the vanishingly small amounts of receptor available in the starting
tissue have hampered biochemical studies.

Recently, six patients with diabetes which could not be alleviated by
treatment with even very large amounts of insulin were found to have
reduced [125] I-insulin binding to their lymphocytes [131]. Sera from 3 of
these women blocked [125] I-insulin binding to cultured human lymphocytes. Only one of these patients had sufficient titer to block [125] I-insulin
receptors in avian erythrocytes. One possibility considered to explain
the lack of apparent antibody activity in the other 3 patients was that
in these patients antibody did not interfere with binding of insulin, but
did alter turnover of receptor. This is a credible possibility, since in MG
antibodies which are not directed at the ACh binding site cause decreases in AChR content [27, 28].

The radioimmune assay devised to detect anti-AChR in the serum of
MG patients used unpurified solubilized AChR labeled with [125] I-toxin
as antigen to detect antibodies which were not directed at the ACh
binding site [15, 16, 17, 113]. Complexes of anti-AChR-AChR-[125] I-αBGT
were precipitated with anti-antibody. By using solubilized insulin receptors labeled with [125] I-insulin as antigen in a similar assay, it should be
possible to detect antibodies to insulin receptors which are not directed
at the binding site. The number of examples of [125] I-hormone labeling of
solubilized receptors is increasing, and radioactive ligands are also
available for other neurotransmitter receptors such as muscarinic
acetylcholine receptors [132] and β-adrenergic receptors [133], among
others. In many of these cases it should be relatively easy to screen
suspect sera for autoantibodies to receptors by this easy approach. This
could lead to the recognition of new autoimmune diseases, provide a
diagnostic test and suggest pathological mechanisms which may be acting
in these diseases.

1.7.3 Prolactin receptors

Antibodies raised against prolactin receptors purified by affinity chromatography from the mammary glands of pregnant rabbits have been used
to test the physiological significance of the purified receptor protein
[134], much as the physiological significance of purified AChR was
demonstrated by showing that antibodies to it blocked AChR activity
in electric organ cells [12]. Unlike antibodies to AChR, antibodies to

the prolactin receptor appear to be directed to the hormone binding site. Antibodies to the prolactin receptor block prolactin binding to both solubilized and membrane bound receptor, thereby preventing its biological effect.

1.7.4 The sodium pump

The sodium pump is a Na^+/K^+-dependent ATPase which translocates Na^+ from inside the cell to outside and the K^+ from outside in. As an integral membrane protein it has much in common with the receptors previously discussed, and is relevant to the discussion here because antibodies have been used quite elegantly in its characterization.

The sodium pump has been purified from dog kidneys [135]. It contains approximately equimolar amounts of two subunits. The larger of the two subunits (135 000 daltons) is known to contain both the binding site for the pump inhibitor ouabain (which is known to act from outside the cell) and the site phosphorylated by ATP (which is known to be on the cytoplasmic side of the membrane) [136]. Hence, at least this subunit spans the width of the membrane. The smaller subunit (40 000 daltons) is a sialo-glycoprotein of unknown function. Carbo-hydrate groups on proteins are generally thought to be located on the outside of the membrane. Rabbits were immunized with purified sodium pump and its subunits. Under saturating conditions these antibodies did not alter the K_m or V_{max} of the enzyme, which may explain why no effect on kidney function was noted in the immunized rabbits [136]. However, ferritin-labeled antibodies were shown by electronmicroscopy to bind to the pump in dog kidney [136]. Curiously, antibodies formed against the high molecular weight chain bound only on the cytoplasmic side of the membrane, whereas antibodies formed against the whole pump bound to both sides of the membrane. Failure of bound antibody to alter pump activity meant that the antibody was not directed to the binding sites for Na^+, K^+ or ATP. But it also meant that transport of ions across the membrane was not achieved by a diffusional carrier mechnism, since that would have involved the energetically impossible task of dragg-ing the bound antibody through the membrane. Instead, some much more subtle conformation change must be involved, perhaps regulating the conformation of a pore through the membrane formed at the inter-face of the subunits [136]. Sodium would be driven out of this pore by phosphorylation of the enzyme and K^+ would pass into the cell during dephosphorylation. It may be that an enzyme like the sodium pump is much less sensitive to perturbation of its structure by binding of antibody than is a receptor like AChR, perhaps because the conformation change between the two states of the pump is strongly driven by phosphorylation of the protein. Transition between the open and closed channel states of AChR, on the other hand, is driven by the minuscule energies involved in reversible binding of the very small ACh molecule. Delicacy of the AChR channel is shown by the fact that the duration and extent of its opening varies with small changes in agonist structure [47], the history of ligand binding [52] (i.e., the receptor desensitized), and even with the membrane potential [50, 51].

1.8 CLOSING COMMENTS

Antibodies to AChR are prime effectors of the autoimmune response in MG and EAMG. Study of EAMG has substantially elucidated the pathological mechanisms in this desease and provided a new diagnostic test for M.G. EAMG is becoming a qualitatively and quantitatively well understood model. A cure for MG clearly must involve some form of immunosuppression. Further EAMG studies should permit progress in this endeavor.

Autoantibodies to other hormone and transmitter receptors are known to be involved in some diseases. Using methods developed for studies of EAMG, autoantibodies may be implicated in other diseases. The pathological mechanisms being discovered in EAMG should prove guides for the study of these diseases.

EAMG provides a method for *in vivo* disturbance of AChR turnover and postsynaptic membrane structure. Study of the disturbed state, like study of a mutation, may provide insight into how the normal state is developed and maintained.

Antibodies to AChR probably have substantial potential yet to be realized as probes for the molecular structure and functioning of AChR.

REFERENCES

1. Lindstrom, J. and Patrick, J. (1974), *Synaptic Transmission and Neuronal Interaction*, M.V.L. Bennett, ed., Raven Press, New York.
2. Olsen, R., Meunier, J.C. and Changeux, J.P. (1972), *FEBS Letters,* **28**, 96–100.
3. Karlin, A. and Cowburn, D. (1973), *Proc. natn. Acad. Sci. U.S.A.,* **70**, 3636–3640.
4. Schmidt, J. and Raftery, M.A. (1973), *Biochemistry,* **12**, 852–856.

5. Eldefrawi, M.E. and Eldefrawi, A.T. (1973), *Arch. Biochem. Biophys.*, **159**, 362–373.

6. Karlsson, E., Heilbronn, E. and Widlund, L. (1972), *FEBS Letters*, **28**, 107–111.

7. Potter, L. (1973), *Drug Receptors*, H.P. Rang, ed., Macmillan, London.

8. Chang, H.W. (1974), *Proc. natn. Acad. Sci. U.S.A.*, **71**, 2113–2117.

9. Biesecker, G. (1973), *Biochemistry*, **12**, 4403–4409.

10. Klett, R.P., Fulpius, B.W., Cooper, D., Smith, M., Reich, E. and Passani, L.D. (1973), *J. Biol. Chem.*, **248**, 6841–6853.

11. Dolly, J.O. and Barnard, E.A. (1975), *FEBS Letters*, **57**, 267–271.

12. Patrick, J., Lindstrom, J., Culp, B. and McMillan, J. (1973), *Proc. natn. Acad. Sci. U.S.A.*, **70**, 3334–3338.

13. Patrick, J. and Lindstrom, J. (1973), *Science*, **180**, 871–872.

14. Lennon, V.A., Lindstrom, J.M. and Seybold, M.E. (1975), *J. exp. Med.*, **141**, 1365–1375.

15. Lindstrom, J., Lennon, V., Seybold, M. and Whittingham, S. (1976), *N.Y. Acad. Sci.*, **274**, 254–274.

16. Lindstrom, J.M., Seybold, M.E., Lennon, V., Whittingham, S. and Duane, D. (1976), *Neurology*, in press.

17. Lindstrom, J. (1976), *J. supramolec. Structure*, **4**, 389–403.

18. Almon, R., Andrew, C. and Appel, S. (1974), *Science*, **186**, 55–57.

19. Appel, S.H., Almon, R.R. and Levy, N. (1975), *New Eng. J. Med.*, **293**, 760–761.

20. Mittag, T., Kornfeld, P., Tormay, A. and Woo, C. (1976) *New Eng. J. Med.*, **294**, 691–694.

21. Aharonov, A., Abramsky, O., Tarrab-Hazdai, R. and Fuchs, S. (1975), *Lancet.* **1**, 340–342.

22. Bender, A., Ringle, S., Engel, W., Daniels, M. and Vogel, Z. (1975), *Lancet*, **1**, 607–609.

23. Bevan, S., Heinemann, S., Lennon, V.A. and Lindstrom, J. (1976), *Nature*, **260**, 438–439.

24. Green, D.P.L., Miledi, R. and Vincent, A. (1975), *Proc. R. Soc. Lond. B*, **189**, 57–68.

25. Elmquist, D., Hoffman, W.W., Kugelberd, J. and Quastel, D.M.J. (1964), *J. Physiol.*, **174**, 417–434.

26. Lambert, E., Lindstrom, J. and Lennon, V. (1976), *N.Y. Acad. Sci.*, **274**, 300–318.

27. Lindstrom, J.M., Einarson, B., Lennon, V.A. and Seybold, M.E. (1976), *J. exp. Med.* **144**, in press.

28. Lindstrom, J.M., Seybold, M.E., Lennon, V.A., Engel, A.G. and Lambert, E.H. (1976), *J. exp. Med.*, **144**, in press.

29. Lindstrom, J. and Lambert, E. (1976), in preparation.

30. Engel, A.G. and Santa, T. (1971), *Ann. N.Y. Acad. Sci.*, **183**, 46–64.

31. Engel, A., Tsujihata, M., Lindstrom, J. and Lennon, V. (1976), *N.Y. Acad. Sci.*, **274**, 60–79.

32. Engel, A., Tsujihata, M., Lambert, E., Lindstrom, J. and Lennon, V., (1976), *J. Neuropath. exptl. Neurol.*, in press.

33. Engel, A.G., Lindstrom, J.M., Lambert, E.H. and Lennon, V.A. (1967), *Neurology,* in press.
34. Illiano, G., Tell, G.P.E., Siegal, M.I. and Cuatrecasas, P. (1973), *Proc. natn. Acad. Sci.,* **70**, 2443–2447.
35. Kohn, L. and Winand, R.J. (1975), *J. Biol. Chem.,* **250**, 6503–6508.
36. Krnjević, K. (1974), *Physiol. Rev.,* **54**, 418–490.
37. Lee, T.P., Kuo, L.F. and Greengard, P. (1972), *Proc. natn. Acad. Sci.,* **69**, 3287–3291.
38. Katz, B. and Miledi, R. (1972), *J. Physiol.,* **224**, 665–699.
39. Fertuck, H.C. and Salpeter, M.M. (1974), *Proc. natn, Acad. Sci.,* **71**, 1376–1378.
40. Miledi, R. (1960), *J. Physiol.,* **151**, 1–23.
41. Berg, D.K., Kelly, R.B., Sargent, P.B., Williamson, P. and Hall, Z.W. (1972), *Proc. natn. Acad. Sci. U.S.A.,* **69**, 147–151.
42. Kuffler, S.W. and Yoshikami, D. (1975), *J. Physiol.,* **251**, 465–482.
43. Heuser, J.E., Reese, T.S., Landis, D.M.D. (1975), The Synapse, *Cold Spring Harbor Symposium XL,* pp. 17–24.
44. Lwebuga-Mukasa, J.S., Lappi, S. and Taylor, P. (1976), *Biochemistry,* **15**, 1425–1434.
45. Hubbard, J.I. and Quastel, D.M. (1973), *Ann. Rev. Pharmacol.,* **13**, 199–216.
46. Cohen, J.B. and Changeux, J.P. (1975), *Ann. Rev. Pharmacol.,* **15**, 83–103.
47. Colquhoun, D., Dionne, V.E., Steinbach, J.H. and Stevens, C.F. (1975), *Nature,* **253**, 204–206.
48. Karlin, A. (1969), *J. gen. Physiol.,* **54**, 245s–264s.
49. Landau, E.M., and Ben-Haim, D. (1974), *Science,* **185**, 944–946.
50. Anderson, C.R. and Stevens, C.F. (1973), *J. Physiol.,* **235**, 655–691.
51. Lester, H.A., Changeux, J.P. and Sheridan, R.E. (1975), *J. gen. Physiol.,* **65**, 797–816.
52. Weber, M., David-Pfeuty, T. and Changeux, J.P. (1975), *Proc. natn. Acad. Sci. U.S.A.,* **72**, 3443–3447.
53. Changeux, J.P., Benedetti, L., Bourgeois, J.P., Brisson, A., Cartand, J., Devaux, P., Grunhagen, H., Moreau, M., Papot, J.L., Sobel, A. and Weber, M. (1975), The Synapse, *Cold Spring Harbor Symposium XL,* pp. 211–230.
54. Bourgeois, J.P., *et al.,* (1972), *FEBS Letters,* **25**, 127-133.
55. Bennett, M.V.L. (1970), *Ann. Rev. Physiol.,* **32**, 471–528.
56. Fambrough, D.M., Drachman, D.B. and Satyamurti, S. (1973), *Science,* **182**, 293–295.
57. Karlin, A. (1974), *Life Sciences,* **14**, 1385–1415.
58. Valderama, R., Weill, C., McNamee, M.G. and Karlin, A. (1976), *Proc. N.Y. Acad. Sci.,* **274**, 108–115.
59. Lee, C.Y. (1972), *Ann. Rev. Pharmacol.,* **12**, 265–286.
60. Lee, C.Y., Tseng, L.F. and Chiu, T.H. (1967), *Nature,* **215**, 1177–1178.
61. Berg, D.K., Kelly, R.B., Sargent, P.B., Williamson, P. and Hall, Z.W. (1972), *Proc. natn. Acad. Sci. U.S.A.,* **69**, 147–151.

62. Heinemann, S. and Lindstrom, J., unpublished.
63. Weill, C., McNamee, M.G. and Karlin, A. (1974), *Biochem. Biophys. Res. Comm.,* **61**, 997–1003.
64. Raftery, M.A., Schmidt, J., Martinez-Carrion, M., Moody, T., Vandlen, R. and Duguid, J. (1973), *J. supramolec. Struc.,* **1**, 360–367.
65. Yogeeswaran, G. and Lindstrom, J., unpublished.
66. Valderama, R., Weill, C., McNamee, M.G. and Karlin, A. (1976), *Proc. N.Y. Acad Sci.,* **274**, 108–115.
67. Cartaud, J., Benedetti, L.L., Cohen, J.B., Meunier, J.C. and Changeux, J.P. (1973) *FEBS Letters,* **33**, 109–113.
68. Meunier, J.C., Sealock, R., Olsen, R. and Changeux, J.P. (1974), *Eur. J. Biochem.* **45**, 371–394.
69. Karlin, A. and Cowburn, D.A. (1973), *Proc. natn. Acad. Sci. U.S.A.,* **70**, 3636–3640.
70. Martinez-Carrion, M., Sator, V. and Raftery, M.A. (1975), *Biochem. Biophys. Res. Comm.,* **65**, 129–137.
71. Hucho, F. and Changeux, J.P. (1973), *FEBS Letters,* **38**, 11–15.
72. Bradley, R.J., Howell, J.H., Romino, W.O., Carl, G.F. and Kemp, G.E. (1976), *Biochem. Biophys. Res. Comm.,* **68**, 557–584.
73. Shamoo, H. and Eldefrawi, M. (1975), *J. Mem. Biol.,* **25**, 47–63.
74. Michaelson, D. and Raftery, M.A. (1974), *Proc. natn. Acad. Sci. U.S.A.,* **71**, 4768–4772.
75. Merlie, J.P., Sobel, A., Changeux, J.P. and Gros, F. (1975), *Proc. natn. Acad. Sci. U.S.A.,* **72**, 4028–4032.
76. Devreotes, P.N. and Fambrough, D.M. (1976), *Proc. natn. Acad. Sci. U.S.A.,* **73**, 161–164.
77. Berg, D.K. and Hall, Z.W. (1975), *J. Physiol.,* **244**, 659–676.
78. Brockes, J.P. and Hall, Z.W. (1975), *Proc. natn. Acad. Sci. U.S.A.,* **72**, 1368–137
79. Berg, D.K. and Hall, Z.W. (1974), *Science,* **184**, 473–475.
80. Devreotes, P.N. and Fambrough, D.M. (1975), *J. Cell Biol.,* **65**, 335–358.
81. Chang, C.C. and Huang, M.C. (1975), *Nature,* **253**, 643–644.
82. Fertuck, H.C., Woodward, W. and Salpeter, M.M. (1975), *J. Cell Biol.,* **66**, 209–2
83. Patrick, J. and Lindstom, J., unpublished.
84. Lindstrom, J., Einarson, B., Francy, M. (1976), *J. supramolec. Struct.,* in press.
85. Katz, B. and Miledi, R. (1973), *Proc. R. Soc. Lond. B.,* **184**, 221–226.
86. Albuquerque, E., Barnard, E., Chin, T.H., Laper, A.J., Dolly O.J., Jansson, S., Daly, J. and Witkop, B. (1973), *Proc. natn. Acad. Sci. U.S.A.,* **70**, 949–953.
87. Lindstrom, J., unpublished.
88. Brockes, J.P. and Hall, Z.W. (1975), 'The Synapse', *Cold Spring Harbor Symposiu* pp. 253–262.
89. Sugiyama, H., Benda, P., Meunier, J. and Changeux, J. (1973), *FEBS Letters,* **35**, 124–128.

90. Heilbronn, E., Mattsson, C., Stalberg, E. and Hilton-Brown, P. (1975), *J. Neurol. Sci.*, **24**, 59–64.

91. Heilbronn, E. and Mattsson, C. (1974), *J. Neurochem.*, **22**, 315–317.

92. Tarrab-Hazdai, R., Aharonov, A., Silman, I., Fuchs, S. and Abramsky, O. (1975), *Nature*, **256**, 128–130.

93. Seybold, M., Lambert, E., Lennon, V. and Lindstrom, J. (1976), *N.Y. Acad. Sci.*, **274**, 275–282.

94. Lennon, V., Lindstrom, J. and Seybold, M. (1976), *N.Y. Acad. Sci.*, **274**, 283–299.

95. Satyamurti, S., Drachman, D. and Stone, F. (1975), *Science*, **187**, 955–957.

96. Sakakibara, H., Engel, A. and Lindstrom, J., (1976) in preparation.

97. Taylor, R.B., Duffus, P.H., Raff, M.C. and de Petris, S. (1971), *Nature*, **233**, 225–229.

98. Heinemann, S., Bevan, S., Kullberg, R., Lindstrom, J., Rice, J. (1976), in preparation.

99. Tarrab-Hazdai, R., Aharonov, A., Abramsky, O. and Yaar, I. (1975), *J. exp. Med.*, **142**, 785–789.

100. Zurn. A. and Fulpius, B.W. (1976), *Clin exp. Immunol.*, **24**, 9–17.

101. Simpson, J. (1960), *Scott. Med. J.*, **5**, 419–436.

102. Strauss, A.J., Segal, B.C., Hsu, K.C., Burkholder, P.M., Nastur, W.L. and Osserman, K.E. (1960), *Proc. Soc. exp. Biol. Med.*, **105**, 184–191.

103. Seybold, M. and Drachman, D. (1974), *New Eng. J. Med.*, **290**, 81–84.

104. Goldstein, G. and Schlesinger, D.H. (1975), *Lancet*, **2**, 256–259.

105. Lennon, V. (1975), *Nature*, **258**, 11–12.

106. Lambert, E.H. and Elmquist, D. (1971), *Ann. N.Y. Acad. Sci.*, **183**, 183–199.

107. Ito, Y., Miledi, R., Molenaar, P.C., Vincent, A., Polak, R.L., von Gilder, M. and Davis, J.N. (1976), *Proc. R. Soc. Lond. B.*, **192**, 475–480.

108. Fambrough, D., Drachman, D. and Satyamurti, S. (1973), *Science*, **182**, 293–295.

109. Albuquerque, E.X., Rash, J.E., Mayer, R.F. and Satterfield, J.R. (1976), *Exp. Neurology*, **51**, 536–563.

110. Abramsky, O., Aharonov, A., Webb, C. and Fuchs, S. (1975), *Clin. exp. Immunol.*, **19**, 11–16.

111. Richman, D.P., Patrick, J., and Arnason, B.G.W. (1976), *New Eng. J. Med.*, **294**, 694–698.

112. Bender, A.N., Ringel, S.P. and Engel, W.K. (1976), *Neurology*, **26**, 447–483.

113. Lindstrom, J. (1976), *Clin. Immun. Immunopath.*, **6**, in press.

114. Toyka, K.V., Drachman, D.B., Pestrouk, A. and Kao, I. (1975), *Science*, **190**, 397–399.

115. Lindstrom, J., Seybold, M., Keesey, J. and Lennon, V. (1976), unpublished.

116. Master, C.L., Dawkins, R.L., Zilko, P.J., Simpson, J.A., Leedman, R.J. and Lindstrom, J. (1976), in preparation.

117. Kao, F. and Drachman, D.B. (1976), *Neurology*, **26**, 383.

118. Aharonov, A., Tarrab-Hazdai, R., Abramsky, O. and Fuchs, S. (1975), *Proc. natn. Acad. Sci. U.S.A.*, **72**, 1456–1459.

119. Abramsky, O., Aharonov, A., Teitelbaum, D. and Fuchs, S. (1975), *Arch. Neurology*, **32**, 684–687.

120. Roitt, J. (1974), *Essential Immunology*, 2nd Edition, Blackwell Scientific Publications, London.

121. Nakamura, R.M. (1974), *Immunopathology*, Little, Brown and Co., Boston.

122. Adams, D.D. and Kennedy, T.H. (1971), *J. clin. Endocr.*, **33**, 47–

123. Dirmikis, S.M., and Munro, D.S. (1975), *Bri. med. J.*, **2**, 665–666.

124. Tate, R.L., Schwartz, H.I., Holmes, J.M., Kohn, L.D., Winand, R.J. (1975), *J. biol. Chem.*, **250**, 6509–6515.

125. Bolonkin, D., Take, R.L., Luber, J.H., Kohn, L.D. and Winand, R.J. (1975), *J. biol. Chem.*, **250**, 6516–6521.

126. Winand, R.J. and Kohn, L.D. (1975), *J. biol. Chem.*, **250**, 6522–6526.

127. Tate, R.L., Holmes, J.M., Kohn, L.D. and Winand, R.J. (1975), *J. biol. Chem.*, **250**, 6527–6533.

128. Mullin, B.R., Fishman, P.H., Lee, G., Aloj, S.M., Ledley, F.D., Winand, R.J., Kohn, L.D. and Brady, R.O. (1976), *Proc. natn. Acad. Sci. U.S.A.*, **73**, 842–846.

129. Illiano, G., Tell, G.P.E., Siegel, M.I. and Cuatrecasas, P. (1973), *Proc. natn. Acad. Sci. U.S.A.*, **70**, 2443–2447.

130. Cuatrecasas, P. (1972), *Proc. natn. Acad. Sci. U.S.A.*, **69**, 318–322.

131. Flier, J.S., Kohn, C.R., Roth, J. and Bar, R.S. (1975), *Science*, **190**, 63–65.

132. Yamamura, H.I. and Snyder, S.H. (1974), *Proc. natn. Acad. Sci. U.S.A.*, **71**, 1725–1729.

133. Wolfe, B.B., Harden, T.K. and Molinoff, P.B. (1976), *Proc. natn. Acad. Sci. U.S.A.* **73**, 1343–1347.

134. Shiu, R.P.C. and Friesen, H.G. (1976), *Science*, **192**, 259–261.

135. Kyte, J. (1971), *Biol. Chem.*, **246**, 4157–4165.

136. Kyte, J. (1974), *J. biol. Chem.*, **249**, 3652–3660.

137. Mattsson, C. and Heilbronn, E. (1975), *J. Neurochem.*, **25**, 899–901.

138. Gage, P.W. (1976), *Physiological Reviews*, **56**, 177–258.

139. Kleinberg, D., Tarrab-Hazdai, R. and Fuchs, S. (1976), *Israel J. Med. Science*, in press.

140. Tarrab-Hazdai, R., Martinez, R.D., Aharonov, A. and Fuchs, S. (1976), *Israel, J. Med. Science*, in press.

141. Lindstrom, J., Dennert, G. and Lennon, V. unpublished.

142. Abramsky, O., Tarrab-Hazdai, R., Aharonov, A. and Fuchs, S. (1976), *J. Immun.* in press.

143. Karlin, A., Holtzman, E., Valderrama, R. and Hsu, K. (1976), *J. Histochem. Cytochem.*, in press.

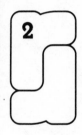

2 Mating-type Interactions in Micro-organisms

MARJORIE CRANDALL *
T.H. Morgan School of Biological Sciences,
University of Kentucky,
Lexington KY 40506

*   Supported by U.S.P.H.S. Grant GM 21889

Receptors and Recognition, Series A, Volume 3
Edited by P. Cuatrecasas and M.F. Greaves
Published in 1977 by Chapman and Hall, 11, New Fetter Lane, London EC4P 4EE
© 1977, Chapman and Hall

INTRODUCTION

This review will focus mainly on the initial phase of microbial mating —
the cell recognition step and the nature of the recognition factors involved in the interaction between complementary sexes. Unfortunately,
the receptor molecules have not been isolated in many systems; in those
systems in which the recognition factors have been characterized
chemically, the nature of the combining sites on the complementary
factors and their interaction are not understood in any detail. Therefore,
this review will also discuss studies on the physiology and biochemistry
of induction of sexual differentiation because knowledge about the
environmental conditions and chemical substances that induce the
synthesis of the recognition factors might yield clues concerning the
nature of these interacting macromolecules and their genetic regulation.

Recent references will be quoted primarily; earlier references including the original report on a particular finding will not always be reitegrated
if available in the bibliography of a quoted article. In Table 2.1 are listed
several review papers on microbial mating systems. This article will
emphasize the literature published since these earlier reviews appeared.
Many authors cited in the text have communicated their recent results to
me prior to publication; these results have contributed greatly to the
development of ideas proposed in this review and their contributions
are greatly appreciated. Attention is called to a forthcoming volume,
Microbial Interactions, (edited by J. Reissig, Biology Department,
C.W. Post College, Long Island University, Greenvale, New York 11548)
to be published in 1977 in Series B of *Receptors and Recognition* in
which certain of the micro-organisms that I have briefly reviewed
will be the subjects of detailed discussions. In addition to reviewing microbial mating interactions; I will also attempt to relate these findings to
mammalian cell interactions wherever possible. The purpose of drawing
attention to parallel observations in unicellular and multicellular systems
is to emphasize the unity of life; 'What is true for *Escherichia coli* is true
for the elephants, except more so.' (Jacques Monod 1954; quoted from
Reissig, 1974). As will be seen in this chapter, many of the biochemical
processes that occur in microbial and mammalian cells are similar even

47

Table 2.1 Reviews on mating in micro-organisms

Group		Type of sexual interaction	Reviews
		Procaryotes	
Bacteria		Transfer of part or all of the chromosome from male to female cell by conjugation	Curtiss, III, (1969); Marvin and Hohn, (1969); Sermonti, (1969); Hopwood *et al.,* (1972); Curtiss, III *et al.,* (1976)
Blue-green algae		Transduction, transformation and cell anastomosis	Padan and Shilo, (1973); Wolk, (1973)
		Eucaryotes	
Algae		Fusion of sexually differentiated haploid gametes forming a diploid zygote	Wiese, (1969); Darden, (1973a); Dring, (1974); Kochert, (1975)
Fungi	Yeasts	Fusion of sexually differentiated haploid cells forming a diploid zygote	Hawker, (1966); Morris, (1966); Crandall and Brock, (1968); Fowell, (1969a,b); Horenstein and Cantino, (1969); Harris and Mitchell, (1973); Biliński *et al.,* (1975); Crandall and Caulton, (1975); Sena *et al.,* (1975); Crandall *et al.* (1976)
	Molds	Hyphal anastomosis or fusion of haploid gametes or gametangia then fusion of haploid nuclei	Cantino, (1966); Esser, (1966); Machlis, (1966); Raper, (1966); Turian, (1966); Bergman *et al.* (1969); Gooday, (1974); Bu'Lock, (1976); Bu'Lock *et al.,* (1976); Horgen, (1976); Sutter (1976); van den Ende, (1976)
Protozoa		Exchange of haploid nuclei between sexually reactive diploid cells	Hiwatashi, (1969); Preer, (1971); Giese, (1973); Miyake, (1974); Sonneborn, (1974a, b)

though they involve quite diverse processes. In some cases, the similarities between different microbial systems and mammalian are so striking as to suggest a fundamental biological law. For example, hormonally-induced differentiation of microbial and mammalian cells occurs only during Gl of the cell division cycle. That specialized protein synthesis occurs only during a gap (resting) period of the cell cycle implies a common control mechanism governing morphogenesis in eucaryotic cells. Such comparisons of studies of normal cells may pave the road for advances in the understanding disease processes and the improvement of health care.

2.1 SEX IN THE INVISIBLE KINGDOM

Many micro-organisms have the ability to reproduce sexually as well as vegetatively. As might be anticipated, the mechanisms of the mating type interactions in different micro-organisms are as diverse as the morphological characteristics separating the taxonomic groups.

In bacteria, the sex of individual cells is determined by cytoplasmic genes located on plasmids. Sexual events in bacteria involve transfer of a DNA fragment from a male donor cell into a female recipient cell. There is no stable diplophase in bacteria; the fragment of donor DNA recombines with the haploid chromosome of the female.

Sexual events in blue-green algae have not been studied extensively but genetic recombination in these procaryotes can be attributed to transduction, transformation or cell anastomosis.

The sex of eucaryotic microbial cells is determined primarily by nuclear genes affecting compatibility; in some cases the expression of these genes is modified by cytoplasmic genes. Sexual events in eucaryotes involve gametic contact between differentiated cells, nuclear fusion, genetic recombination during meiosis with the production of new haploid cells (or nuclei) as part of the regular alternation of haploid and diploid phases of the life cycle. These types of sexual interactions are listed in Table 2.1 together with earlier reviews in each area.

Differences in mating-type interactions between bacteria and eucaryotes are contrasted in Fig. 2.1 and Fig. 2.2. Despite these differences, all of the mating systems share the following general characteristics. (1) Differentiation of vegetative cells into sexually reactive cells is determined genetically but may be induced by substances secreted by the opposite sex or by various environmental factors such as nutritional limitation, metal ions, exogenous chemicals or light.

(a) The F-pili Conduction Model
(after Brinton, 1965)

Male pilus determined by F factor

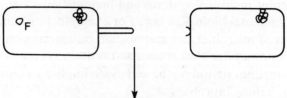

Cell recognition between tip of F pilus
and female receptor site

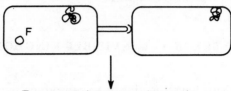

Transfer of chromosome from male
to female through the pilus

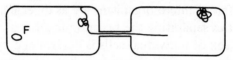

(b) The F-pili Retraction Model
(after Marvin and Hohn, 1969, and Curtiss, III, 1969;
figure redrawn from Bradley, 1972)

Cell recognition between tip of F pilus
and female receptor site

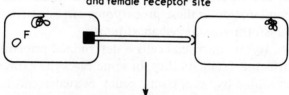

Retraction of pilus by depolymerization
mechanism

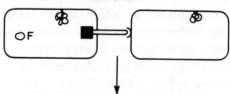

Conjugation bridge formation and
transfer of chromosome

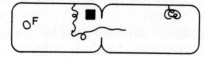

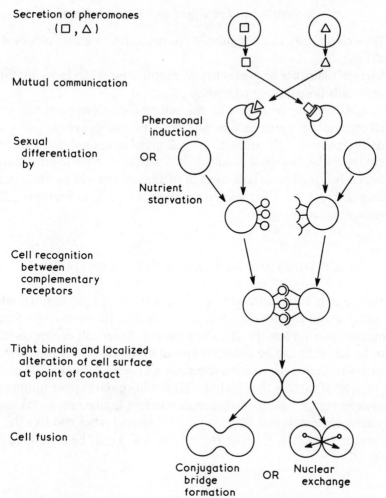

Secretion of pheromones
(□ , △)

Mutual communication

Pheromonal
induction

Sexual
differentiation
by   OR

Nutrient
starvation

Cell recognition
between
complementary
receptors

Tight binding and localized
alteration of cell surface
at point of contact

Cell fusion

Conjugation   OR   Nuclear
bridge   exchange
formation

Fig. 2.2 Recognition steps during conjugation in eucaryotic micro-organisms. The early steps in recognition between opposite sexes in eucaryotic microbial systems are represented here and discussed in detail in the text. The salient features illustrated are that cells are stimulated to differentiate into gametes as a result of pheromonal induction or nutrient starvation. Then recognition between opposite sexes occurs via complementary receptors located on surface filaments. Tighter binding between cells is effected by fusion of the surface layers of the cell.

Fig. 2.1 Illustration of two models for bacterial conjugation. Both the F-pili conduction and retraction models for chromosome transfer from male to female bacteria are based on substantial observations and both may be correct. This idea is discussed more fully in Section 2.3.1).

(2) The cell division cycle regulates commitment to either conjugation or mitosis.

(3) Surface filaments or organelles of motility establish loose binding between cells preparatory to mating.

(4) Complementary receptors on the cell surfaces determine the specificity of cell recognition between opposite sexes or mating types.

(5) At the point of cell contact, the cell membrane (and cell wall, if present) must be dissolved to allow for transfer of genetic information or for cell fusion. The details of each of these steps will be discussed in Sections 2.2 and 2.3.1 for bacterial conjugation and in Sections 2.2–2.5 for eucaryotic conjugation.

2.2 INDUCTION OF SEXUAL DIFFERENTIATION

Vegetative cells will continue to grow and divide as long as nutrients are available and the opposite sex is unavailable. However, following communication via sex-specific chemicals or direct cell contact with the opposite sex, cells can be diverted into mating activities. Commitment to the pathway that leads to gametogenesis and cell fusion may be governed by a variety of factors that include: (i) sex hormones (pheromones); (ii) environmental conditions (such as nutrient limitation, metal ions, exogenous chemicals and light); (iii) cytoplasmic genes and (iv) the stage in the cell cycle. Each of these factors will now be examined in detail.

2.2.1 Sex hormones (pheromones) inducing mating events

Micro-organisms produce a multitude of chemical substances that affect different stages in the process of mating: (i) gametogenesis, (ii) chemotaxis or chemotropism of the differentiated gametes, (iii) cell fusion (conjugation) or (iv) later sexual events (nuclear fusion and meiosis). Most of these chemicals are sex-specific in that they are secreted only by one mating type and act only on the other mating type. However, some substances from a clonal culture act on the secreter cells themselves to cause selfing or differentiation of a somatic cell into a germ cell; others stimulate both sexes non-specifically while in still other cases, substances from unrelated genera stimulate sexual events in the organism under study.

Sex-specific chemicals are variously called sex hormones, sex factors, gamones, erogens or are named after the genus or species of the producing micro-organism or the target organ affected. In this review, microbial substances that affect sexual events will be referred to as pheromones (from the Greek, pherein, to carry, plus horman, to excite; Karlson and Luscher, 1959).

Some pheromones are secreted into the growth medium and are isolated and purified from culture filtrates; others are volatile and can, therefore, act at a distance; while still others are exchanged only by direct cell contact. Only a few pheromones have been characterized chemically; these were found to be either lipids, peptides, organic acid derivatives or glycoproteins. However, most pheromones have been identified only by their biological activity. When isolated pheromones are added to tester cells, a series of steps is initiated that results in the production of germ cells from somatic cells in algae; conjugation or sexual sporulation in fungi; or cell adhesion, meiosis and fertilization in protozoa. Usually the assay used to measure the pheromonal levels focuses only on the end result of such multi-step processes. In actual fact, the steps affected by pheromones are probably very early in sexual morphogenesis. However, in no case has the site of action of these substances been discovered. Similarly, no information is available concerning the attachment of pheromones to the cell wall or membrane. Reissig (1974) has reviewed mating type interactions in micro-organisms with a view toward understanding this problem of how the microbial cell surface decodes regulatory signals transmitted by pheromones or cell contact. He has proposed a useful term — perisemic (from the Greek: peri, all around; and sema, signal) to mean any process by which a regulatory signal is transduced along or across the cell envelope. Reissig cites many instances, in addition to mating-type interactions, in which a substance secreted by one microbe induces a morphological change in another cell type. One such example of a perisemic process is the interaction of the peptide pheromone α-factor from the yeast *Saccharomyces cerevisiae* with the opposite mating type (strain **a**). α-Factor induces (i) Gl arrest; (ii) localized wall synthesis; (iii) secretion of the complementary peptide pheromone and (iv) sexual agglutination (reviewed in Grandall *et al.*, 1976). How can one molecule initiate such diverse events? Reissig (1974) discusses these processes in terms of membrane effects and offers[some stimulating ideas for future study.

Not all pheromones are synthesized constitutively. In some cases, one mating type produces a pheromone constitutuvely that induces the opposite type to synthesize another pheromone which in turn acts upon

Table 2.2 Pheromonal inducers of gametogenesis

Pheromonal production*	Pheromone	Organism	References
		Algae	
C	Glycoprotein from ♂ colonies	Volvox aureus	Reviewed in Darden, (1970, 1973a); Kocherts, (1975)
C	Glycoprotein from ♂ colonies	Volvox carteri	Kochert and Yates, (1974); Pall, (1974); Starr and Jaenicke, (1974)
		Fungi	
C–I	Antheridiol = hormone A (sterol) from ♀; hormone B from ♂	Achlya ambisexualis	Gooday, (1974); Horgen and Ball, (1974); Horowitz and Russell, (1974); Horgen, (1976)
C–C	Antheridiol = hormone A (sterol) and oogoniol = hormone B (sterol) from hermaphroditic (⚥) mold	Achlya heterosexualis	Barksdale, (1969); Barksdale and Lasure, (1974); McMorris et al., (1975)
I–I	Trisporic acids [oxidized, unsaturated derivatives of 1, 1, 3-trimethyl-2-(3'-methyloctyl) cyclohexane] produced from mating type-specific precursors synthesized by (+) and (−) strains	Blakeslea trispora	van den Ende et al., (1972); Sutter et al., (1973); Bu'Lock et al., (1974)
		Mucor mucedo	Mesland et al., (1974); Wurtz and Jochusch, (1975)
		Phycomyces blakesleeanus	Bergman et al., (1969); Sutter, (1975; 1976)

Table 2.2 Pheromonal inducers of gametogenesis (*continued*)

Pheromonal production*	Pheromone	Organism	References
C–I	Sex factors: α-factors (peptides) from strain α and a-factor I (unknown) from strain a produced constitutively; a-factor II (peptide) from strain a produced inducibly	*Saccharomyces cerevisiae*	Levi, (1956); Duntze et al., (1970); Duntze et al., (1973); Duntze (1974); MacKay and Manney, (1974); Wilkinson and Pringle (1974); V. MacKay, (personal communication); reviewed in Crandall et al., (1976)
		Protozoa	
C–I	Gamone I [glycoprotein (2 × 10⁴ daltons)] in mating type I: Gamone II [calcium –3– (2′-formylamino-5′-hydroxybenzoyl) lactate] in mating type II	*Blepharisma intermedium*	Miyake and Beyer, (1973); Braun and Miyake, (1975)

* C = Constitutive; I = Inducible with respect to pheromonal synthesis or secretion (see Section 2.2.1, for explanation).

the first cell type to induce further sexual changes. In other systems, pheromonal synthesis is apparently induced by cell contact because no soluble substances are detectable until both mating types are mixed. Thus a classification of sexual systems may be devised according to the type of pheromonal synthesis by both mating types: constitutive – constitutive (C–C), constitutive–inducible (C–I), or inducible–inducible (I–I). The pheromonal class is indicated in Table 2.2 and discussed in individual examples below.

Gametogenesis

Soluble factors (pheromones) are known in many microbial systems that induce vegetative cells to differentiate into gametes. In general, systems in which gametogenesis is induced by pheromones are distinct from systems in which gametogenesis is induced by nutrient limitation (compare Tables 2.2 and 2.4). Pheromonal inducers of gametogenesis have been identified in the three major groups of eucaryotic protists: algae, fungi and protozoa.

Gametogenesis in algae

The green alga, *Volvox aureus* is homothallic and dioecious (both male and female colonies differentiate from the same clonal culture). Differentiated male colonies constitutively produce a glycoprotein that induces vegetative colonies to produce sperm (reviewed in Darden, 1973a; Kochert, 1975). Darden (1973a) suggests that this glycoprotein might be inhibiting female differentiation rather than acting to induce male organs. Similar findings were obtained with the species *V. carteri* which is heterothallic (diclonal). The male spheroids produce a glycoprotein inducer but this acts in two ways: it causes the female strain to produce eggs and the male strain to produce sperm (reviewed in Darden, 1973a; Kochert, 1975). Filtrates from several other species of *Volvox* (*V. rousseletii, gigas, obversus* and *dissipatrix*) also contain sexual inducers. These inducers are: (i) species-specific; (ii) large molecules (10^4 to 2×10^5 daltons) that readily aggregate; (iii) relatively heat stable; (iv) resistant to treatment with DNase, RNase, trypsin or chymotrypsin; (v) destroyed by pronase; (vi) conjugated to carbohydrate and (vii) active at extremely low concentrations (reviewed in Kochert, 1975). The glycoprotein inducing substance from *V. carteri* has been purified to near homogeneity (Kochert and Yates, 1974). It consists of 60% protein and 40% sugar (Starr and Jaenicke, 1974). Molecular weight estimates vary between 2.5 and 3.2×10^4 daltons depending upon the technique employed. The content of aspartic acid was about 9% of the amino acid residues; threonine about 8% and serine 7 to 11% (kochert and Yates, 1974; Starr and Jaenicke, 1974). The main sugars were xylose (26% of the carbohydrate content) and glucose (33%) (Starr and Jaenicke, 1974). The inducer binds to the saccharide binding site of concanavalin A and is eluted from a column of this ligand with glucose (Pall, 1974). Thus the male inducer has been characterized chemically but its mechanism of action remains unknown. In addition to this glycoprotein inducer of male differentiation, an RNA-histone complex has been found to stimulate gamete production in *Volvox aureus*. Darden (1973b) reported that a phenol extract from male colonies will stimulate sperm production when arginine-rich histone is mixed with the extract. The histone apparently complexes with RNA which is the predominant component of this extract because an increase in turbidity occurs when the extract and histone solutions are mixed. This complex could be the inducer or, alternatively, the extract may be contaminated with traces of the glycoprotein male inducing substance. Darden discusses these observations in terms of induction of gametogenesis versus the removal of an inhibitor of gametogenesis.

Gametogenesis in fungi

A large number of fungi produce pheromones that induce sexual differentiation. The identification and chemical characterization of these fungal pheromones is a very active area of research and many current reviews of this subject are available (Gooday, 1974; Bu'Lock, 1976; Bu'Lock *et al.*, 1976; Crandall *et al.*, 1976; Horgen, 1976; Sutter, 1976; van den Ende, 1976). The reader is referred to these reviews and earlier reports cited therein for detailed discussions of each biological system and the chemical structures of the pheromones produced. Only brief highlights of the latest results will be presented in the following text.

Pheromones synthesized by fungi generally: (i) are of relatively low molecular weight; (ii) are either lipids or peptides; (iii) are sex- and species-specific; (iv) diffuse through the air or liquid medium to the opposite sexes or sex organs; (v) cause gametic differentiation in the opposite type; (vi) function at very low concentrations probably via gradients and (vii) are, in some cases, destroyed by the target (recipient) cell (Gooday, 1974; Bu'Lock, 1976). Lipid inducers of fungal gametogenesis include sterols from *Achlya* and trisporic acids from the Mucorales; peptide inducers include a-factor and α-factor from *Saccharomyces*. There are other fungal pheromones known that fall into these two classes of chemical compounds but they will be discussed in later sections on chemotaxis, chemotropism and cell fusion.

Both the female and the male organs of the water mold *Achlya* produce sterol pheromones that cause vegetative mycelia to differentiate into gametangia (reviewed in Horgan, 1976). Production of male sex organs (antheridial branches) in *Achlya ambisexualis* is induced by a sterol pheromone called hormone A (antheridiol) that is produced constitutively by the female (Barksdale, 1969). Antheridiol induces the male to produce another sterol pheromone called hormone B (oogoniol) (Barksdale and Lasure, 1974; McMorris *et al.,* 1975). Hormone B induces the female to produce sex organs (oogonial branches). Therefore, the heterothallic species, *Achlya ambisexualis,* can be described as C−I with respect to pheromonal production. On the other hand, the hermaphroditic species, *Achlya heterosexualis,* produces both antheridiol and oogoniol constitutively (McMorris *et al.,* 1975).

Within 30 min after the addition of antheridiol to vegetative mycelium, synthesis of poly A-rich RNA is increased. After 3 h, a stimulation of protein synthesis is observed. After 3 to 4 h, differentiated male sex organs are observed (reviewed in Horgen, 1976). Differentiation is inhibited by actinomycin D, therefore, RNA synthesis is required (Horowitz and Russell, 1974). Acetylation of histone-like nuclear proteins is associated with antheridiol-stimulated differentiation (Horgan and Ball, 1974). This acetylation of chromosomal proteins may be the mechanism by which the steroid pheromone alters the pattern of gene expression by allowing transcription of genes involved in gamete production. Because antheridiol and oogoniol affect reproductive processes they may be viewed as plant counterparts to androgens and estrogens (McMorris *et al.,* 1975). It is not known how the *Achlya* steroid pheromones act to alter gene expression but in higher eucaryotes steroid hormones act by complexing with a cytoplasmic receptor protein; the hormone-receptor complex then enters the nucleus and presumably stimulates transcription (O'Malley and Means, 1974).

Pheromonal induction of sexuality has been described in other fungi: *Ascobolus stercorarius* (Bistis and Raper, 1963; *Glomerella cingulata* (Driver and Wheeler, 1955); *Bombardia lunata* (Zickler, 1952); *Sapromyces reinschii* (Bishop, 1940); *Thraustotheca primoachlya* and *Dichtyuchus achlyoides* (Salvin, 1942) and *Dictyuchus monosporus* (Sherwood, 1966) but in none of these systems have the sexual substances been characterized

In contrast, the sexual substances produced by the mucoraceous fungi have been the subject of many chemical studies (reviewed in Sutter, 1976; Bu'Lock *et al.,* 1976; van den Ende, 1976). The best studied genera of the Mucorales are *Blakeslea, Mucor* and *Phycomyces.* These fungi have a common mechanism of sexuality that involves trisporic acid pheromone synthesized from mating type-specific precursors. Both the plus and minus partners contribute to the synthesis of trisporic acids (van den Ende *et al.,* 1972; Bu'Lock *et al.,* 1974). Therefore, these fungi may be assigned to the I–I class of pheromonal production because the rate of synthesis of trisporic acids is quite low in single strains but is markedly increased in mixed cultures of plus and minus. Trisporic acids induce sexual differentiation in both mating types and can be considered as intrahyphal rather than interhyphal regulators of sexual development. This conclusion is based on studies of mutants with abnormal carotene (*car*) synthesis that exhibit aberrant sexual behavior. These *car* mutants do not stimulate the wild type strains to produce zygophores (sexually specialized branches of the mycelium) but zygophores are formed on the mutant indicating that the mutant does not produce the trisporic acid precursor for the opposite type (Sutter, 1975). Mutant studies by Wurtz and Jockusch (1975) confirm this hypothesis that uptake of trisporic acids is not necessary for sexual interaction.

In wild type strains, the trisporic acid that is induced stimulates: (i) carotogenesis; (ii) synthesis of higher levels of mating type-specific trisporic acid precursors and (iii) zygophore development in both strains (Sutter, 1975, 1976). Interestingly, the trisporic acid precursors synthesized by plus and minus mating types are volatile and can induce zygophores when both sexes are separated by an air space (Mesland *et al.,* 1974; Sutter 1976). The mating type-specific precursors from plus strains are called prohormones and are designated P^+; these are converted into trisporic acids by minus cultures. Prohormones from minus strains are designated P^- and are converted into trisporic acids by plus cultures. Restated, neither P^+ nor P^- are precursors of each other but both are precursors of trisporic acids. Both P^+ and P^- are complex mixtures of oxygenated derivatives of C_{15} or C_{18} trisporic acid molecules (Bu'Lock *et al.,* 1976). These workers have proposed a scheme for the synthesis of trisporic acids consistent with the mating type-specific conversion steps.

While this scheme qualitatively explains how each mating type produces prohormones which the other can convert into trisporic acids, it does not quantitatively account for the high levels of trisporic acids produced in mated cultures. Only 0.2 mg l^{-1} of P^+ or P^- are found in plus or minus cultures but up to 200 mg l^{-1} of trisporic acids are found in mixed plus and minus cultures. These results may be explained if the end product induces higher levels of the precursors.

Several hormone-like factors have been described for the *Saccharomyc* *cerevisia* mating system (reviewed Crandall *et al.,* 1976). These sex-specific pheromones include: (i) **a** hormone and α-hormone that cause cell volume expansion; (ii) **a**-factor I, **a**-factor II (peptides and proteins) and α-factors (a family of peptides) that cause Gl arrest, cell wall changes (shmoos) and induce agglutinability; (iii) α substance-I that induces agglutinability (probably identical to α-factor); (iv) an inhibitor of α-factor action, and (v) diffusible factors that cause sex-specific budding responses. Cell-cell interactions during mating in yeast will be reviewed by T.R. Manney and J. Meade in *Microbial Interactions.*

Gametogenesis in protozoa
In general, sexual competence in protozoa is triggered by nutrient limitation of a matude culture (one that has undergone many divisions after conjugation). But, in some cases the heterologous cell-free medium reduces the time for induction of cell interaction. Soluble factors in the medium induce selfing in *Euplotes patella* (Kimball, 1939) and induce competence for mating in *Tetrahymena pyriformis* (Phillips, 1971), *Blepharisma intermedium* (Miyake and Beyer, 1973), and *Oxytrichia bifaria* (Ricci *et al.,* 1975). However, the substances responsible for inducing the cells to conjugate have been identified only in *Blepharisma* (Miyake and Beyer, 1973; Braun and Miyake, 1975). The chemical composition of the two gamones from *Blepharisma* are given in Table 2.2 Gamone I stimulates mating type II to secrete gamone II which in turn stimulates mating type I to unite with mating type II in pairs (Miyake and Beyer, 1973; Miyake, 1974). Therefore this mating system may be classified as constitutive-inducible with respect to pheromone synthesis. The mode of action of gamone I as an agglutinin is discussed in Section 2 The mode of action of gamone II as possibly a calcium ionophore is discussed in Section 2.2.2. In addition, both gamone I and II act as chemotactic agents (see below). Gamones are secreted by other species of *Blepharisma* also. A study of species specificity of these pheromones is useful in establishing phylogenetic relationships within genera (Miyake and Bleymann, 1976). This is a good example of how the study of sexual pheromones can lead to cross-fertilization between unrelated disciplines.

Chemotaxis or chemotropism
Gametes of many micro-organisms are motile and are attracted toward the opposite sex chemotactically. Non-motile gametes can bend or grow toward the opposite sex. This latter response is referred to as

chemotropism. A variety of substances secreted by cells can act as chemotactic agents for gametes; many are normal metabolites that are secreted by both sexes and, therefore, are not sex-specific. In these cases, the motile cells are probably responding to the concentration gradient which increases with proximity to the secreter (reviewed in Darden, 1973a). If only one sex is motile (e.g., the sperm) or if only one sex has the chemotactic receptor, then specificity is built into the system even though both cells secrete the same substance. The chemotactic agents that are sex-specific and that have been characterized chemically are listed in Table 2.3.

As a rule, sex attractants are produced after exponential growth has ceased. Therefore, these agents are considered secondary metabolites (after Bu'Lock, 1967). In fact, the entire process of sexual differentiation may be viewed as secondary metabolism (see Section 2.2.2) because sexual processes usually occur only during the portion of the G1 stage of the cell cycle when the cell is not growing (see Section 2.2.3).

Although much is known about the chemistry of the sex-specific attractants, in no case have the sensory receptors on the cell surface for these agents been identified. However, from many studies of sex-, species- and generic-specificity of chemotactic agents in algae, it was found that the receptors respond to a variety of related lipid compounds (reviewed in Dring, 1974). For example, the chemotactic agent from *Chlamydomonas moewusii* var. *rotunda* is volatile and its effect can be mimicked by coal gas, ethylene and ethane (Tsubo, 1961). It, therefore, appears that lipid-receptor interactions lack the degree of specificity inherent in protein recognition complexes.

Chemotaxis of algal gametes
Many cases of chemotactic responses of motile gametes in algae are known (reviewed in Darden, 1973a; Dring, 1974; Gooday, 1975). But only a few chemotactic attractants have been characterized chemically (see Table 2.3); all the chemotactic agents from algae are volatile lipids. [Lipids also function as chemotropic agents in fungi (see below)]. In addition, it is possible that the glycoprotein agglutinins of *Chlamydomonas* may also function to attract gametes since streamers of agglutinin material are observed to diffuse out from the flagellar tips of gametes (Brown *et al.*, 1968). [Likewise, the agglutinin from *Blepharisma* also functions to attract the opposite sex (Honda and Miyake, 1975; see below)]. The three algal attractants that have been characterized

Table 2.3 Chemotactic agents that attract motile gametes

Substances	Organism	References
Algae		
Multifidene (C_{11} polyene, $C_{11}H_{16}$); volatile	*Cutleria multifida*	Müller, (1974); Jaenicke *et al.*, (1974)
Ectocarpen [allo-*cis*-1-(cyclohepta-2′, 5′-dienyl) but-l-ene, $C_{11}H_{16}$]; volatile	*Ectocarpus siliculosus*	Müller *et al.*, (1971); Müller, (1972); Jaenicke and Müller, (1973)
Fucoserraten (1, 3-*trans*,-5-*cis*-octatriene, C_8H_{12}); volatile	*Fucus serratus*	Jaenicke and Seferiadis, (1975)
Fungi		
Antheridiol and oogoniol (see Table 2.2)	*Achlya* spp.	Barksdale, (1969); Barksdale and Lasure, (1974); McMorris *et al.*, (1975)
Sirenin (bicyclic sesquiter-penediol = $C_{15}H_{24}O_2$) produced by ♀ gametangia, attracts motile ♂ gametes	*Allomyces arbuscula*	Nutting *et al.*, (1968); Barksdale, (1969)
Protozoa		
Gamone I attracts type II cells; Gamone II attracts type I cells (see Table 2.2)	*Blepharisma intermedium*	Honda and Miyake, (1975)

(multifidene, ectocarpen and fucoserraten, see Table 2.3) are low molecular weight, volatile lipids that are secreted by the female and attract the motile male gametes (their structures and biosynthesis are reviewed in Gooday, 1975). Fucoserraten is produced by *Fucus serratus* (Jaenicke and Seferiadis, 1975). A related species, *F. vesiculosus* produces the sex attractant *n*-hexane (Hlubucek *et al.*, 1970). The sperms are apparently sensitive to a concentration gradient rather than to *n*-hexane in particular. The attractants from these two closely related organisms have not been compared for cross-specificity.

Sexual chemotaxis and chemotropism in fungi
The only fungi that have motile gametes are the Phycomycetes; among these, *Allomyces* is the only example where a chemotactic agent (sirenin) has been characterized (see Table 2.3). Sirenin is a lipid, as are the chemotactic agents in algae. A striking characteristic of the fungal sex attractants is that they are active at extremely low concentrations (10^{-8} to 10^{-11} M; Gooday, 1974). Fungal pheromones cause positive chemotropic growth responses between non-motile gametes or thalli. For example, many of the fungal pheromones that induce sexual differentiation (Table 2.2) also cause the differentiated mycelial structures to grow in the direction of the opposite sex. The chemotropic fungal pheromones include antheridiol and oogoniol (both sterols) from *Achlya* and trisporic acids (also lipids) from the Mucorales. These attractants cause the gametangia or zygophores to grow toward one another and then fertilization occurs (reviewed in Gooday, 1975; Bu'Lock, 1976).

In the case of the single-celled fungi, yeast produce protuberances in liquid medium in response to pheromones isolated from the opposite mating type. When opposite mating type yeast cells are placed close together but not touching on agar, one or both cells elongate toward each other. Such sex-specific conjugation tube formation has been reported in: *Saccharomyces cerevisiae* (Levi, 1956; Duntze *et al.*, 1970; MacKay and Manney, 1974; see Table 2.2), *Tremella mesenterica* (Bandoni, 1965; Reid, 1974) and other *Tremella* species, *Rhodosporidium toruloides* (Abe *et al.*, 1975) and *Sirobasidium magnum* (Flegel, personal communication). In yeasts that do not secrete pheromones (e.g., *Hansenula wingei*, *Schizosaccharomyces pombe* and *Ustilago violacea*; see Table 2.4), copulatory processes are not elaborated until direct contact between cells or between surface filaments is established.

The ascomycetous yeast *Saccharomyces cerevisiae* produces peptide pheromones that cause shmoo-shaped cells in the opposite mating type (discussed above; see Table 2.2). The heterobasidiomycetous yeast *Tremella mesenterica* produces amino acid or peptide 'erogenes' constitutively that induce conjugation tubes in the opposite mating type (Reid, 1974). The molecular weights of these substances are less than 1000 daltons; they can be separated into three components by silica gel or paper chromatography. Mating type A of the yeast *Rhodosporidium toruloides* (imperfect form called *Rhodotorula glutinis*) constitutively synthesizes an inducing substance that causes mating type *a* to form conjugation tubes and to secrete *a* factor that causes A cells to form conjugation tubes (Abe *et al.,* 1975). Just as in the case of *Saccharomyces cerevisiae, a* and A factors inhibit bud formation in the complementary mating type. The heterobasidiomycete *Sirobasidium magnum* (Flegel, 1976) produces conjugation tube stimulators (of between 7000 and 13 000 daltons) only if opposite types are separated by membrane tubing; no pheromones are isolatable from single or mixed type cultures (Flegel, personal communication). These findings indicate that each mating type may inactivate or inhibit the pheromone from the opposite type. Similar observations of inactivation of the pheromone from the opposite sex have been reported for antheridiol and sirenin (reviewed in Gooday, 1974) and for α-factor (Hicks and Herskowitz, 1976).

In addition to the sex-specific shmoo formation in *Saccharomyces cerevisiae,* a sex-specific budding response was reported in this species (Herman, 1971a). Cells of complementary mating type of *Kluyveromyces lactis, Hansenula anomala* and *Hansenula wingei* also exhibited inter-generic budding responses indicating that the diffusible constituents possess broad specificity.

Sexual chemotropism has been reported in a wide variety of other fungi no mentioned here but the substances involved have not been characterized (reviewed in Gooday, 1975).

Sexual chemotaxis and chemotropism in protozoa
Just as in fungi, there are motile and non-motile mating types in protozoa that produce chemotactic and chemotropic sex attractants, respectively. In the case of the motile ciliate *Blepharisma intermedium,* both gamone I and gamone II (see Table 2.2) cause sexual differentiation, promote the production of the complementary gamone and attract the opposite mating type chemotactically (Honda and Miyake, 1975). However, in the sessile ciliate *Tokophrya infusionum,* the mating reaction is initiated by elongation of opposite mating type cells toward each other (Sonneborn, 1963).

Cell fusion

Macrocysts represent the sexual stage (zygote) of the myxomycete *Dictyostelium discoideum*; they form as the result of cell fusion of many amoeboid cells (Clark *et al.*, 1973; Erdos *et al.*, 1973). Sexual hormones secreted by cellular slime molds determine commitment to macrocyst production. Aggregated amoebae that have begun to form a fruiting body will form a macrocyst instead if phereomones extracted from other strains are added (Filosa *et al.*, 1975). There are homothallic (self-compatible) strains that produce macrocysts in clonal cultures but most strains produce macrocysts only when paired (Erdos *et al.*, 1973). However, the significance of the pairing is not clear; the strains do not appear to be heterothallic in the strict sense of the term because macrocysts form in only of the paired strains when the partners is separated by dialysis tubing (O'Day and Lewis, 1975; MacHac and Bonner, 1975). Therefore, it would be more appropriate to call these pairs secreters and responders (Lewis and O'Day, 1976) rather than opposite mating types.

The pheromone from *D. discoideum* strain NC-4 is about 12 000 daltons (O'Day and Lewis, 1975) and may be volatile (D. O'Day, personal communication). The pheromones from *D. purpureum* secreter strains Dp6 or Dp7 are less than 12 000 daltons and heat labile (Lewis and O'Day, 1976). The pheromone from *D. mucoroides* homothallic strain Dm7 is water-soluble and volatile (Filosa *et al.*, 1975).

There is also evidence for pheromones influencing cell fusion and later conjugal events in the ciliate *Blepharisma intermedium*. In addition to gamone I and II (Miyake and Beyer, 1973; Miyake, 1974) which are involved in the initial mating (agglutination) reaction (see Table 2.2), there are also pheromones transmitted from one cell to the next during gametic contact (Miyake, 1975). Multicellular complexes of type II cells agglutinated by gamone I will remain united without undergoing cell furion. However, if one cell of complementary type I fuses with a cell at the end of the row of agglutinated cells of type II, then nuclear cycles of sexual differentiation begin at the site of union and are propagated sequentially through the homotypic complex. This suggests that the nuclear processes of conjugation in protozoa are controlled by intracellular pheromones or signals that can be transmitted through the membrane from one cell to the next.

Cellular communication between normal liver cells that are touching in culture can be measured by the transmission of pulses of electric current or the dye fluorescein from one cell to the next (Loewenstein, 1972). Communicating junctions form rapidly where cell membranes come into contact. The calcium ion concentration on both sides of the membrane plays an important role in the permeability transformations associated with the making and breaking of junctions between these mammalian cells. (See Section 2.2.2) for a discussion of the role of calcium ion in cell fusion in protozoan cells).

2.2.2 Environmental factors affecting mating

Many micro-organisms have evolved elaborate mechanisms, such as the secretion of pheromones and sex attractants (see Section 2.2.1), to ensure that sex will be an intimate part of their lives. However, other microbes leave the fate of their sexual activities to chance occurrences of conditions favorable for mating (i.e., the absence of nutrients, the presence of certain metal ions or chemicals or the proper amount of light). In general, sexual reproduction occurs within a narrower range of environmental conditions than vegetative growth. An extensive review of many environmental influences (nutrient concentration, carbon and nitrogen sources, mineral elements, growth factors, water supply, pH, temperature, aeration, carbon dioxide, volatile substances, light, gravity, inhibitors, contact stimuli, wounding) on fungal reproduction by Hawker (1966) revealed that virtually nothing was known about the molecular basis for the effects of these different factors. Now, ten years later, some progress has been made to identify the biochemical events involved in microbial mating but it is clear that despite the wealth of publications in this area, there is still a dearth of knowledge and understanding about such basic biological process as sexual differentiation. Science marches on, slowly.

Nutrient limitation
Sexual differentiation of vegetative cells into gametes in many eucaryotic microbial systems is mediated by depletion of an essential nutrient rather than by pheromones (discussed in Section 2.2.1). In fact, it appears that these two modes of induction of sexual differentiation in

eucaryotes are mutually exclusive; in most systems in which gametogenesis is induced by nutrient limitation, no evidence for the existence
of pheromones has been found (where sought). A comparison of
Tables 2.2 and 2.4 will demonstrate this point; organisms that produce
pheromones can mate under conditions favorable to growth whereas
organisms that do no produce sex pheromones become sexually
reactive under growth-limiting conditions. However, as more research is
done, this generalization may be proved incorrect since Phillips (1971)
has found that while induction of mating reactivity in *Tetrahymena
pyriformis* requires starvation, soluble factors are secreted into the
medium under these conditions that induce mating between unreactive
cells of the two mating types. Also, Barksdale (1962) and Mullins and
Warren (1975) found that while induction of sexuality in *Achlya* is
controlled hormonally, starvation works in concert to produce promote
sexual morphogenesis.

Sexual morphogenesis induced by starvation may be considered a
survival mechanism because it allows the cell to produce either a
resistant spore that does not divide or another cell better able to survive.
For example, mating functions are not present during growth in ciliates
but as growth ceases the mating substances responsible for agglutination
appear on the cilia (Metz, 1954). If conjugation or autogamy do not
occur, these protozoan cells will become senescent and die (Preer Jr.,
1971; Miyake, 1974). If mating does occur, the cells are rejuvenated.
Furthermore, survival of exconjugants increases even more if cells are
starved for another 1 to 2 days after the peak of sexual reactivity
(Vaughan and Barnett, 1973).

In general, nitrogen starvation triggers sexual morphogenesis
(Table 2.4). When a nitrogen source, such as a required amino acid, is
withheld from *Chlamydomonas,* cells differentiate into gametes without
any net protein synthesis. Under these conditions, DNA and RNA
synthesis and protein turnover continue at the same rates as in vegetative
cells but tRNA is synthesized preferentially (Jones *et al.,* 1968). Apparently enough of the required amino acid is generated by breakdown
of vegetative proteins to synthesize conjugation-specific proteins.
Similarly, auxotrophic mutants of *Hansenula wingei* conjugate in the
absence of their required amino acids even though protein synthesis is
required for mating (Crandall and Brock, 1968). In some microbial

Table 2.4 Induction of mating reactivity by nutrient limitation

Nutrient conditions	Organism	References
	Algae	
Nitrogen starvation	*Chlamydomonas reinhardti*	Kates and Jones, (1964); Jones *et al.*, (1968); Schmeisser (1973)
Nitrogen starvation or Phosphorus starvation	*Golenkinia minutissima*	Ellis and Machlis, (1968)
Nitrogen starvation	*Oedogonium cardiacum*	Hill and Machlis, (1970)
Sulfur starvation	*Pandorina unicocca*	Rayburn, (1974)
Nitrogen starvation	*Scenedesmus obliquus*	Trainor and Burg, (1965)
	Fungi	
Nitrogen starvation	*Achlya ambisexualis*	Barksdale, (1962); Mullins and Warren, (1975)
Nutrient limitation	*Candida lipolytica*	Herman, (1971b)
Salt solution	*Dictyostelium discoideum*	Clark *et al.*, (1973)
Phosphate starvation	*Dictyostelium discoideum*	Erdos *et al.*, (1973)
Vitamin limitation or nutrient limitation	*Hansenula wingei*	Reviewed in Crandall and Caulton, (1975)
Nutrient limitation	*Phialophora dermatitidis*	Ouyezdsky and Szaniszlo, (1973)
Nitrogen starvation	*Schizosaccharomyces pombe*	Reviewed in Crandall *et al.*, (1976)
Nutrient limitation	*Ustilago violacea*	Day, (1972)
	Protozoa	
Nutrient limitation	*Paramecium*	Reviewed in Hiwatashi, (1969)

systemes, either sulfur or phosphorus limitation can induce sexuality. Since an energy source is necessary for macromolecular synthesis during gametogenesis, it is difficult to study the effect of carbon source limitation in heterotrophs (such as fungi and protozoa). However, with photoautotrophs (such as algae), cells will form gametes in distilled water as long as light is available as an energy source (see below) and CO_2 (or bicarbonate buffer) is supplied as a carbon source.

In bacteria, the conditions that yield high mating reactivity are just the opposite of that which is found with the higher protists. A rich medium is better than a synthetic medium for donor cultures. Starvation of the male (in buffered-saline or for a required amino acid) results in loss of pili, loss of the ability to form specific pairs with the female and an increase in recipient ability (F^- phenocopies). Donor cells grown anaerobically (in broth without shaking) prior to mating have more F pili per cent and transfer their chromosomes more effectively (Curtiss III, *et al.*, 1969).

Metal ions and exogenous chemicals

Increasing attention is being given to the functions of various metal ions in stimulating microbial mating. While the mechanism of action of different metals in promoting conjugation is not understood, it is tempting to speculate on the following possibilities.

(1) Any cation may act as a counterion, neutralizing the negative charges on the cell surfaces that prevent agglutination in distilled water.

(2) Ca^{2+} may be involved specifically in promoting cell contact by forming bridges between sugars on opposing cell surfaces. For example, α-fucose is a common terminal sugar of oligosaccharide chains on glycoproteins; Ca^{2+} is chelated by a pair of hydroxyl groups from each of two fucose molecules and is hydrated by three water molecules to form a bridge. Such a bridge could act in cross-linking cells (Cook and Bugg, 1975).

(3) Mn^{2+} may activate cell surface glycosyltransferases.

(4) K^+ may be required for translocation of the recognition factors across the membranes (Takahashi and Hiwatashi, 1974).

(5) The intracellular K^+/Ca^{2+} ratio may affect membrane depolarization and, hence, cell fusion. Membrane depolarization is correlated with chemical induction of mating which produces a low intracellular concentration of Ca^{2+} relative to K^+ (Cronkite, 1976). Interestingly, the

Table 2.5 Induction of sexuality by metal ions or exogenous chemicals

Conditions	Organism	References
	Bacteria	
Pretreatment of females with Zn^{2+}	Escherichia coli	Ou, (1973); Ou and Reim, (1976)
	Algae	
Low concentrations of EDTA	Chlamydomonas	Wiese and Jones, (1963)
Mn^{2+}-poor	Ditylum brightwellii	Steele, (1965)
	Fungi	
Mn^{2+}	Aspergillus nidulans	Zonneveld, (1975)
Ca^{2+}, Mg^{2+}-poor, or EDTA or V	Hansenula wingei	Reviewed in Crandall and Caulton, (1975)
O_2 + ergosterol	Kluyveromyces lactis	Mas et al., (1974)
β-Sitosterol	Phytophthora spp.	Elliott, (1972); Ribeiro et al., (1975)
Cholesterol, β-sitosterol, stigmasterol, progesterone	Pythium spp.	Al-Hassan and Fergus, (1969); Hendrix, (1970); Child and Haskins, (1971)
Cu-containing pheromone	Saccharomyces cerevisiae	Duntze et al., (1973)
Testosterone, estradiol	Saccharomyces cerevisiae	Yanagishima, (1971)
O_2 (shift from anaerobic to aerobic); addition of peptone or casamino acids	Schizosaccharomyces pombe	Calleja, (1973)
	Protozoa	
Ca^{2+}-containing pheromone	Blepharisma intermedium	Kubota et al., (1973)
Ca^{2+}-poor + KCl or EDTA or Mn^{2+} or Co^{2+} or additions of acriflavin, proflavine, heparin, acetamide or biuret	Paramecium spp.	Reviewed in Hiwatashi, (1969)

conjugation-initiating pheromone from another ciliate, *Blepharisma*, chelates one calcium ion in a complex with three pheromone molecules and two waters of hydration (Kubota *et al.*, 1973). This pheromone in its free form (not complexed with calcium) may function as an ionophore to carry calcium ions outside the cell thereby affecting membrane depolarization and mating reactivity. An analogous situation obtains in another animal system; the calcium ionophore A23187 decreases the internal Ca^{2+} concentration in echinoderm eggs and activates fertilization events (Steinhardt and Epel, 1974).

(6) Certain metal ions such as Ca^{2+} or Mg^{2+} that are normally present on the vegetative cell surface as structural components may inhibit cell fusion; these may be removed by EDTA which promotes sexual differentiation in many micro-organisms (Wiese and Jones, 1963; Hiwatashi, 1969; Crandell and Caulton 1975).

(7) Similarily, these normal metal ions that are inhibitory to conjugation may be replaced by analogs such as Zn^{2+}, V^{4+} (= VO^{2+}) or V^{5+} (= VO_3^-), Cu^{2+}, Co^{2+}, Mn^{2+} or other trace metals that could compete for the sites occupied by Ca^{2+} or Mg^{2+} by virtue of their similarities in electronic structure.

In addition to metal ions, chelating agents and metal ion-containing pheromones (ionophores), certain other chemicals, including steroids, induce mating in micro-organisms (summarized in Table 2.5).

Light

The mating systems of some eucaryotic micro-organisms are light-sensitive; either requiring light as an energy source (as in algae) or responding to the daily (circadian) cycle of light and dark. In other cases, light inhibits sexual activities (Table 2.6).

2.2.3 Cell cycle regulation of mating

In all microbial mating systems studied in details, sexual reproduction occurs best under conditions restricting growth. Growth-limiting conditions include pheromonally-induced Gl arrest (see Section 2.2.1), nutrient starvation (see Section 2.2.2), stationary phase of growth, excess or limited metal ion concentrations or presence of chemical inhibitors (summarized in Tables 2.4 and 2.5). Under these conditions, cells are not

Table 2.6 Variation in mating response with light

Conditions	Organism	References
	Algae	
Light required as an energy source for gametic differentiation	*Chlamydomonas reinhardti*	Kates and Jones, (1964)
Blue light at 450 nm induces sexuality	*Golenkinia minutissima*	Ellis and Machlis, (1968)
	Fungi	
Formation of macrocyst (diploid zygote) inhibited by light	*Dictyostelium discoideum*	Clark *et al.*, (1973); Erdos *et al.*, (1973)
Ooospore formation inhibited by light	*Phytophthora parasitica*	Honour and Tsao, (1974)
	Protozoa	
Peak of sexual agglutinability at dusk; nadir at dawn	*Euplotes crassus*	Miyake and Nobili, (1974)
Diurnal appearance of mating competency during the light cycle (daytime)	*Paramecium bursaria*	Cohen, (1965); Imafuku, (1975)
Diurnal change in mating type	*Paramecium multimicronucleatum*	Reviewed in Hiwatashi, (1969); Clark, (1972)

synthesizing DNA and are arrested at a point in the cell cycle there is an even number of chromosomal complements (G1) (summarized in Table 2.7). If cells were to mate during the S stage in the cell cycle, then the zygotes would not have an even number of chromosome sets. There is apparently a point of commitment within G1 where the cell can either initiate differentiation (by synthesizing certain specific proteins) or enter another division cycle (by synthesizing more DNA) (Wolfe, 1976).

While the cell cycle of procaryotes is not defined in the same terms, similar events occur during bacterial mating; cell division in the female is prevented because DNA synthesis is inhibited by contact with the male (Skurray and Reeves, 1973).

Differentiation or cell fusion in higher eucaryotic cells also occurs at Gl. Terminal differentiation of mammary epithelial cells requires hormone-dependent events that can only occur during a limited portion of the Gl phase. Insulin (a peptide hormone) induces these cells to produce milk proteins but there is a critical stage at mid Gl when cells are sensitive to insulin. If cells pass through this stage and arrest in late Gl, then they must undergo another division, passing through S, G2, M and early Gl arrest, in order to become sensitive again to hormonal induction of differentiation (Vonderhaar and Topper, 1974). Fusion of myoblast cells to form multinucleate skeletal muscle fibres only occurs in the Gl stage. Interestingly, those cells capable of fusing are not synthesizing DNA but do synthesize myosin and actin which can be detected on the cell surface by antibody binding assays. On the other hand, myoblast cells in S, G2 or M cannot fuse (perhaps due to their altered surface properties) and they do not bind antibody against myosin (Okazaki and Holtzer, 1966). Thus it can be concluded that the synthesis of certain cell surface proteins or secreted proteins involved in very specific functions unrelated to growth occurs only when the cell is not synthesizing more DNA, RNA and protein for the purpose of cell proliferation.

In other words, growth or primary metabolism is incompatible with differentiation or secondary metabolism. In discussing the regulation of secondary biosynthesis, Bu'Lock, (1967) states that exhaustion of nutrients in the stationary phase of growth disrupts the primary metabolic pathway and may lead to the production of high levels of intermediary metabolites normally present at only low concentrations. These intermediates may then induce the synthesis of new enzymes or enzymes already present may now act on these substrates previously

Table 2.7 Induction of sexuality during the cell division cycle

Stage in cell cycle at which mating occurs	Organism	References
	Algae	
G1	Chlamydomonas reinhardii	Schmeisser et al., (1973)
	Fungi	
Stationary phase (G1)	Hansenula wingei	Crandall and Brock, (1968)
G1	Saccharomyces cerevisiae	Sena et al., (1975); reviewed in Crandell et al., (1976)
G1	Schizosaccharomyces pombe	Streiblova and Wolf, (1975)
G1	Sirobasidium magnum	T. Flegel, (personal communication)
G1 for mating type a_1; all stages for mating type a_2	Ustilago violacea	Cummins and Day, (1973); Poon et al., (1974); Cummins and Day, (1974)
	Protozoa	
G1 micronucleus and a G1 macronucleus	Diophrys scutum	Luporini and Dini, (1975)
G1 micronucleus and a G1 macronucleus	Oxytrichia bifaria	Ricci et al., (1975); Luporini and Dini, (1975)
G1 macronucleus and a G2 micronucleus	Tetrahymena pyriformis	Wolfe, (1974; 1976); Doerder and Debault, (1975)

limited in concentration because of competing primary processes.
Bu'Lock asks the question: 'Why should the detailed course of secondary metabolism vary so spectacularly from one organism to another when the primary processes are so similar? Secondary metabolites are an expression of the individuality of the species in molecular terms, and possibly these are the most appropriate terms in which that individuality can be expressed, since the 'senses' of micro-organisms are ... chemical. By their fruits they shall be known ...'.

2.3 ROLE OF SURFACE FILAMENTS IN CONJUGATION

'Many of the physical properties which have been determined as of the cell surface belong more to the extraneous coatings than to the actual protoplasmic surface of the cell.' (Robert Chambers, 1940; quoted by Revel and Ito, 1967).

Surface specializations consisting of fine filaments extending from the plasmalemma are observed on many types of animal cells (ex., amoebae, interstinal epithelium). These filaments are referred to as the fuzzy coat and are thought to play a functional role in adsorption of particles (Revel and Ito, 1967).

In many microbial systems, there are filamentous appendages on the cells of one or both sexes. These surface filaments extend into the medium and are the first part of the cell to make contact with the opposite sex. In all instances, it is presumed that the recognition factors (see Section 2.4) are located on these filaments and that the massive agglutination reaction that occurs when opposite cell types are mixed is due to overlapping of filaments and complex formation between complementary recognition factors located on the surfaces of the appendages.

In bacteria, the filament involved in conjugation, the sex pilus, is found only on the male. In algae, the recognition factors are located on the flagella tips. In fungi, there are long filaments called fimbriae and shorter filaments referred to as hairs or 'fuzz' that may be involved in cellular recognition. In protozoa, the recognition factors are located on the ventral cilia. In all of these systems, the surface filaments or organelles of motility, as extensions of the cell, facilitate the initial binding together of the gametes (summarized in Table 2.8).

Table 2.8 Surface filaments implicated in conjugation

Filament	Organism	References
	Bacteria	
Pili	*Escherichia coli*	Brinton, (1965), Curtiss III, (1969)
	Pseudomonas	Mayer, (1971); Bradley, (1972)
	Algae	
Flagella	*Chlamydomonas*	Wiese, (1974)
	Fungi	
Surface filaments	*Hansenula wingei*	Reviewed in Crandall and Caulton, (1975); reviewed in Crandall *et al.*, (1976)
Surface fuzz	*Saccharomyces cerevisiae*	Osumi *et al.*, (1974)
Surface filaments	*Schizosaccharomyces pombe*	Yoo *et al.*, (1971); Calleja *et al.*, (1976)
Fimbriae	*Ustilago violacea*	Day and Poon, (1975)
	Protozoa	
Cilia	*Blepharisma*	Miyake and Honda, (1976)
	Paramecium	Miyake, (1964); Takahashi *et al.*, (1974)

2.3.1 Bacterial pili

In Fig. 2.1, two models for bacterial conjugation are depicted. The F-pili conduction model, Brinton, (1965) assumes that the DNA from the male is injected into the female through the empty channel inside the F pilus. The F-pili retraction model proposed by Marvin and Hohn, (1969 and independently by Curtiss III, (1969) also recognizes the importance of the F pilus in the initial stages of conjugation, i.e., cell recognition. But it is proposed that following contact with a recipient, the pilus is retracted by a depolymerization mechanism at its base thereby drawing the two cells close together. The depolymerization mechanism is activated by a signal initiated by contact of the recipient cell with the tip of the pilus on the donor. The chromosome is not transferred until a cell wall conjugation bridge is formed between intimately-joined couples. The retraction model is an agreement with early observations of conjugation in *Escherichia coli* (Anderson, 1958) in which conjugation is seen in the light microscope to proceed first by the formation of tentative connections between cells by 'invisible' fibres followed by a close joining of the two bacteria. Evidence compatible with the pili retraction theory has been found in *Pseudomonas* also (Mayer, 1971; Bradley, 1972). Evidence in support of the pili conduction model is presented by Ov and Anderson (1970). These two models on bacterial conjugation are re-evaluated by M. Achtman and R. Skurray in a review entitled, 'A redefinition of the mating phenomena in bacteria' for *Receptors and Recognition,* Series B, Microbial Interactions (ed. J. Reissig). In another recent review by Curtiss III, *et al.,* (1976), both models of bacterial conjugation are considered carefully and it is concluded that, in fact, both may be correct. To explain this situation these authors state: 'That *Escherichia coli,* and presumably other conjugationally proficient organisms, should have two means to accomplish genetic transfer should not be too surprising since backup and/or alternate mechanisms for accomplishing a necessary objective are standard fare in biological organisms, even with regard to sexuality (Comfort, 1972)'.

2.3.2 Algal flagella

In *Chlamydomonas,* the recognition factors are known to be located distally on the flagella as evidenced by the following observations;
(1) Initial gametic contact occurs only at the flagellar tips of each pair and,
(2) When agglutinin from one sex is added to cells of the opposite sex the cells clump together, joined by their flagella, and form rosettes (Wiese 1974).
As would be anticipated, isolated flagella bind specifically to cells of the opposite mating type (reviewed in Wise, 1969). Binding of flagella labelled with a radioactive isotope to unlabelled cells can serve as a useful assay in the study of the recognition factors (Snell, 1976a). Membrane vesicles (called isoagglutinins) containing the recognition factors apparently bleb off from the flagella during growth and may be recovered from the culture medium (see Section 2.4).

2.3.3 Fungal fimbriae

Surface hairs or fuzz that may carry the recognition factors are found on the surfaces of mating types of the yeasts *Saccharomyces cerevisia* (Osumi *et al.,* 1974), *Schizosaccharomyces pombe* (Yoo *et al.,* 1971;

Calleja *et al.*, 1976, and *Hansenula wingei* (A. Day, electron micrographs published in Crandall and Caulton, 1975). During the early stages of agglutination, the surface filaments of *H. wingei* overlap forming a region of greater density between the cells (K. Aufderheide, personal communication; S. Conti, electron micrograph published in Crandall *et al.*, 1976). In this latter micrograph, the cell walls are seen to be distorted along the line of contact even though the cell walls are not yet touching. This observation indicates the great strength of the attractive forces between the cell surfaces.

2.3.4 Protozoan cilia

The recognition factors responsible for the mating reaction in *Paramecium* are located on the ventral cilia. Cell-free cilia preparations of one mating type can cause sex-specific agglutination of cells of the complementary type (Miyake, 1964) or the cilia can aggregate with isolated cilia from the complementary type (Takahashi *et al.*, 1974). Membrane vesicles containing the recognition factors may be solubilized from isolated cilia (see Section 2.4). D. L. Nanney will be reviewing 'Cell–cell interactions in ciliated protozoa' for *Microbial Interactions.*

2.4 COMPLEMENTARY RECEPTORS INVOLVED IN CELL RECOGNITION

Biological specificity results from the interaction of complementary molecular structures (Pauling, 1974). In micro-organisms, the specificity of initial cell contact between opposite sexes is determined by complementary recognition factors on the respective cell surfaces. That recognition factors are, in fact, complementary has been demonstrated in studies of the sexual agglutination factors responsible for the massive cell clumping reaction that precedes conjugation in the yeast *Hansenula wingei* (Crandall and Brock, 1968). When the agglutination factors are extracted from each cell type and mixed together, they form a soluble, neutralized complex (Crandall *et al.*, 1974); this report is the first demonstration that isolated recognition factors from microbial mating systems interact in a manner analogous to antibody–antigen complex formation. Complementarity between two molecules is determined by their ability to bind specifically to each other and not to other molecules from similar sources. Macromolecular binding depends on the formation of many weak chemical bonds, the most important of which are hydrogen bonds (Pauling, 1974). Hydrogen bonds are involved in the sexual agglutination reaction of *H. wingei* since the cell clumps are disrupted by autoclaving in 8 M urea (Crandall and Brock, 1968).

The agglutination factors from mating types *5* and *21* of *H. wingei* are called *5*-factor (*5f*) and *21*-factor (*21f*), respectively. Both *5f* and *21f* may be located on surface filaments that extend from the cell walls (evidence presented in Crandall and Caulton, 1975; Crandall *et al.*, 1976). These agglutination factors may be solubilized by preparation of cytoplasmic extracts or by enzymatic digestion of whole cells. Both mating factors are mannanproteins (Crandall and Brock, 1968). Solubilized *5f* is multivalent and causes sex-specific agglutination of strain *21* cells whereas *21f* is univalent and must be assayed by its specific inhibition of *5f* agglutination activity.

Criteria for determining whether these isolated macromolecules are the mating type-specific receptors are presented in Crandall and Brock (1968):

(1) *5f* is obtained only from mating type *5*; *21f* is obtained only from mating type *21*.

(2) *21f* is released from the cell surface by the same treatment (trypsin digestion) that inactivates agglutination of strain *21* cells.

(3) *5f* specifically agglutinates strain *21* cells but not cells of strain *5* or the *5* x *21* hybrid.

(4) *21f* specifically inhibits *5f* agglutination of strain *21* cells.

(5) Each mating factor is adsorbed only to the cell surface of the opposite mating type and not to its respective cell type or to the diploid hybrid.

(6) Cells with a large amount of adsorbed factor from the complementary type are inhibited from agglutinating with the complementary type whereas treatment of the respective cell type with the same concentration of mating factor does not interfere with cell association with the opposite type.

(7) The isolated mating factors neutralize each other *in vitro* by forming a *5f–21f* complex.

(8) The diploid hybrid does not produce either mating factor or the neutralized complex and is non-agglutinative with either mating type.

Cytoplasmic extracts of strain *5* cell yield *5f* molecules of different sizes (Brock, 1965), with the number of combining sites roughly proportional to molecular weight (Crandall *et al.*, 1974). On the other hand, subtilisin digestion of strain *5* cells yields a homogeneous

5-agglutinin molecule (about 10^6 daltons) that has 6 combining sites (Taylor and Orton 1968). Small binding fragments liberated from the large central core by reduction were estimated to be 12 000 daltons by Taylor and Orton (1968). Yen and Ballou (1974) found these binding fragments to be composed of 28 amino acids and 60 mannose units (corresponding to a molecular weight of 12 500 daltons). The central core of *5f* is 10% protein and 90% mannose which is present in chains of 8 sugar units linked to serine or threonine. These two hydroxy amino acids constitute 62% of the core protein. The molecule also contains 5% phosphate and, hence, should be called a phosphomannanprotein (Yen and Ballou, 1974). However, recent studies of the chemical nature of wild type and mutant 5-agglutinin lacking phosphate indicates that the phosphate is not required for the agglutination reaction (Sing *et al.*, 1976). The *5f* remains active after boiling for 5 min (Brock, 1965); this heat stability of the *5f* may be explained by a lack of α-helical structure in this branched, randomly-coiled molecule (Taylor and Tobin, 1966). The standard free energy of association of *5f* with strain *21* cells is -14.5 kcal mol^{-1} which is high for reversible reactions and is due to co-operativity between the several combining sites on the *5f* molecule (Taylor and Orton, 1970, 1971).

Cytoplasmic extracts or trypsin digestion of strain *21* cells yields *21f* molecules that are univalent and homogenous with respect to molecular weight (about 40 000 daltons) (Crandall and Brock, 1968; Crandall *et al.*, 1976). The *21f* contains 25 to 35% carbohydrate (mainly mannose) and is unstable to protein denaturants but stable to reducing agents.

Isolated *5f* and *21f* neutralize each other by forming a complex that is soluble probably because *21f* is univalent and cannot form crosslinks as in the case of precipitating antibodies. The *5f–21f* complex is stable presumably because of the large negative free energy of association (reported by Taylor and Orton, 1970). The neutralized complex is detected using a biological assay developed by Crandall and Brock, (1968) in which *21f* is destroyed by alkali and *5f* agglutination activity is recovered. Using this assay, it was found that the *5f–21f* complex exists in multiple forms reflecting the molecular weight heterogeneity of the cytoplasmic *5f* used to form the complex. Three peaks of complex were detected with estimated molecular weights of 0.5, 1.2 and 3.8 x 10^6 daltons the corresponding number of combining sites on the *5f* for these peaks were calculated to be roughly 6, 16 and 63 (Crandall *et al.*, 1974). The cytoplasmic *5f* studied by Brock, Crandall and co-workers can vary in size from 15 000 to 10^8 daltons whereas the cell surface *5f* studied by

Taylor and co-workers and Ballou and co-workers is released by subtilisin and is homogeneous (10^6 daltons). It seems reasonable that heterogeneous cytoplasmic *5f* may represent both precursor molecules and aggregates thereof which will become cell surface *5*-agglutinin.

There is no enzymatic destruction of either factor during complex formation since the complex is stable and can be purified using conventional methods in protein chemistry. Furthermore, the activity of one of the factors can be recovered from the complex following specific chemical inactivation of the complementary factor. No attempts have been made yet to determine whether either factor has enzymatic activity.

Activity of the 5-agglutinin is destroyed by both pronase and exo-α-mannanase (Yen and Ballou, 1974) but it is not known at this time whether the specificity. of interaction of the agglutinin with *21f* is due to protein–protein or carbohydrate–protein interaction. Yen and Ballou (1974) presented additional results that, together with the results of enzymatic inactivation, were consistent with the suggestion that the specificity of binding of *5f* to strain *21* cells resides in the protein moiety but that the mannan serves to maintain the structural configuration of the *5f* required for agglutination activity. It might be possible to define further the functional relationships of the protein and mannan portions by recovering the products of enzymatic digestion of *5*-agglutinin and using these small fragments of peptide or polysaccharide as specific inhibitors of complex formation between the two solubilized agglutination factors (R. Laine, personal communication). Inhibition of *5f–21f* complex formation by *5f* fragments may be a more sensitive assay than inhibition of *5f* agglutination of *21* cells since the *21* cells have many binding sites that can still complex *5f* even though some sites were covered up with binding fragments from *5f*. In agreement with this idea, attempts to obtain fragments of *5*-agglutinin by enzymatic digestion that would still inhibit *5f* agglutination of strain *21* cells were unsuccessful (C.E. Ballou, personal communication).

The mating receptors on male and female bacteria have not been characterized. Only *Escherichia coli* has been studied in detail and, in this bacterium, the female receptor site appears to involve both the carbohydrate portion of the cell wall lipopolysaccharide plus an outer membrane protein (references cited in Table 2.9). Little is known about the male receptor site except that it is located at the tip of the F pilus (Ou, 1975; reviewed in Curtiss III, *et al.,* 1976). Antigenic differences between the ends and the shafts of sex pili have been discovered but these differences were not described in relation to the male mating receptor (Meynell *et al.*, 1974).

Mating type-specific recognition substances, called isoagglutinins or gamones, may be recovered from the culture medium of differentiated gametes of *Chlamydomonas*. These preparations are large molecular weight particles (10^8 daltons) that contain glycoproteins (Wiese and Hayward, 1972). The (−) isoagglutinin of *C. eugametos* contains (in molar percentage) 12% aspartic acid, 9% threonine and 9% serine; that of *C. moewusii* syngen II contains 8% aspartic acid, 9% threonine and 10% serine. Both isoagglutinins contain rhamnose, mannose, galactose, arabinose, xylose and several more as yet unknown components (Wiese, 1974). Recently several laboratories have discovered that gamone preparations are membrane vesicles with mastigonemes (flagellar appendages) attached (McLean *et al.*, 1974; Bergman *et al.*, 1975; Snell, 1976b). Purified mastigonemes are not active in agglutination whereas membrane vesicles (after removal of the mastigonemes) are extremely active as agglutinins, indicating a direct involvement of the membrane in the mating reaction. Presumably, the agglutination factors on the surface of these membrane vesicles are the same as the recognition factors located at the tips of the flagellar membrane of the gametes (see Section 2.3.2).

The flagellar mating type substances of *Chlamydomonas moewusii* are susceptible to different enzymes *in situ*. The (−) agglutinin is sensitive to proteases whereas the active portion of the (+) agglutinin appears to be a carbohydrate in which an α-glycosidically bound mannose is in a functionally-essential terminal position (Wiese, 1974). Conflicting results were obtained by McLean and Brown Jr., (1974) who reported that trypsin eliminates the mating ability of only the (+) gamete. It is possible that the conditions for trypsin treatment were different in these two laboratories and as a result, different phenomena were observed. For example, Crandall and Brock, (1968) reported that trypsin treatment of strain *21* at high cell density caused the release of sufficient *21f* for purification but the cells were still agglutinative. On the other hand, trypsin treatment of strain *21* at low cell density caused inactivation of agglutinability but not enough *21f* was released into the supernatant to be detectable.

The results of Wiese (1974) in which a protein on the (−) gamete was found to interact with a terminal, α-glycosidically-bound mannose on the (+) gamete is the first demonstration that mating type association cab be mediated by protein–carbohydrate binding. Roth *et al.*, (1971) proposed that the complementary macromolecules responsible for adhesion in mammalian cell culture may be glycosyltransferases on one cell and the acceptor substrate on the opposing cell. If this were true for microbial mating system, then the following predictions must be fulfilled:

(1) The mating type locus should code for two complementary agglutination factors that exhibit glycosyltransferase and endogenous polysaccharide acceptor activity, respectively;

(2) Agglutinated gametes as well as the *in vitro* complex between the recognition factors should dissociate in the presence of the required sugar nucleotide plus the proper co-factors for the glycosyltransferase;

(3) Sterile mutants should show loss of glycosyltransferase or acceptor activity.

The possibility that the recognition factors can also function in the synthesis of the conjugation tube is attractive and merits further investigation. Some preliminary evidence favoring this hypothesis has been reported by McLean and Bosman (1975) in studies with *Chlamydomonas*. Glycosyltransferase activities for galactose, glucose, N-acetylglucosamine, N-acetylneuraminic acid, mannose and fucose are present at higher levels in gametes than in vegetative cells and when the (+) and (−) gametes are mixed, there is an increase in these activities. There is no such increase when (+) and (−) vegetative cells are mixed. Flagellar membrane vesicles from (+) and (−) gametes have high levels of transferase ecto-enzyme activities and these also yield enhanced activity upon mixing. While these activities are correlated with gametic function, there is no evidence that they are correlated with recognition function. Further studies comparing agglutination and transferase activities in parallel as a function of time under different conditions are required. For example, McLean and Bosmann (1975) point out that their results might be related simply to the threshold number of available acceptor sites on the surface of gametes or vesicles. Therefore, the role of these glycosyltransferase enzymes in recognition is still uncertain.

Homotypic cell pairing is induced in mating type II of *Blepharisma intermedium* by blepharmone, a gamone from mating type I. Blepharmone is a glycoprotein of 20 000 molecular weight with 5% carbohydrate that consists of 3 residues of glucosamine plus 3 residues of mannose per gamone molecule. It has the typically high amounts of aspartic acid (26%), serine (20%) and threonine (16%) characteristic of glycoproteins but, in addition, has an unusually high content of tyrosine (13%) (Braun and Miyake, 1975). Thus blepharmone resembles 5-agglutinin from *Hansenula wingei* which also (i) causes homotypic agglutination of the opposite sex; (ii) is found in the culture fluid; (iii) is a glycoprotein containing mannose and (iv) contains a high content of serine (53%) and threonine (9%) (Yen and Ballou, 1973).

Both gamones I and II induce an increase in protein synthesis in the complementary mating type and, subsequently, homotypic pairing via ciliary binding. However, the proteins induced by gamone II appear to be the same as the normal components of the mating type I cell surface; no conjugation-specific protein was found (Miyake and Honda, 1976). Therefore, these workers postulate that the function of blepharmone is to induce the production of greater quantities of a receptor normally present on the cell surface. Mating type I receptors are particles of 10^6 daltons or larger and are associated with lipid. Evidence is presented by Miyake and Honda, (1976) that suggests the induced receptors accumulate at the antero-ventral region of the cell surface where union takes place; if gamone is removed, the cells separate whereas if cyclo-heximide is added (after union is established) the cells remain united indicating that previously synthesized proteins can aggregate at special-ized regions of the cell surface where they complex with the complement-ary recognition factor. Ricci *et al.*, (1976) report that mating type I receptor for gamone II is distinct from membrane receptor for ligands such as phytohemagglutinin, concanavalin A and antitubulin antibodies. Yet these latter ligands inhibit gamone II activity perhaps by restricting the mobility of membrane components.

Cilia isolated from sexually reactive *Paramecium caudatum* and treated with urea plus EDTA, release particles that retain the ability to cause sex-specific agglutination of reactive tester cells of complementary mating type (Kitamura and Hiwatashi, 1976). These urea–EDTA particles are membrane vesicles (0.1 μm in diameter) which contain at least 20 proteins and 4 kinds of glycoproteins. The proteins extracted from ciliary vesicles from both mating types were similar in size distribution and, therefore, no mating type-specific proteins were detected in the two different preparations. Such differences could have been masked if the two recognition factors were of the same molecular weight. Attempts to resolve these components further by sonication of the vesicles resulted in loss of agglutination activity.

Membrane receptors are also involved in cell fusion in the myxomycete *Didymium iridis.* The zygote plasma membrane exhibits a protein not found on the plasma membrane of individual amoeba before fusion. This surface membrane protein is of high molecular weight (J. Yemma, personal communication).

In the absence of methods for the assay, isolation and purification of the mating type-specific recognition factors, it is tempting to perform studies on whole cells in the search for mating type-specific differences. However, this approach is fraught with potential pitfalls since in most systems the two opposite mating type strains being studied are not isogenic. Therefore, a difference in a surface property between the two sexes may be due to a gene product unrelated to the mating type locus. In the absence of detailed genetic and biochemical analyses of the recognition factors and the second molecule, it is not possible to relate differences in surface properties of gametes to a sex-specific function. However, it is worthwhile mentioning a few examples of surface differences that have been reported.

Certain reagents such as concanavalin A and Alcian Blue bind to glycoproteins (Wardi and Allen, 1972) and, therefore, can be used in studies of the cell surface. However, these reagents may not bind to the recognition factors but rather to other unrelated substances. For example, Wiese and Hayward (1972) found that concanavalin A binds to the flagellar tips of both sexes of *Chlamydomonas reinhardti* but did not prevent agglutination. Therefore, in this species, the binding sites for concanavalin A are different than the sexual contact sites (also on the flagellar tips). However, different results were obtained with another species, *C. moewusii*, in which (tetravalent) concanavalin A inactivates the (+) sex specifically (Wiese and Shoemaker, 1970). The identity of the concanavalin A binding site and sexual receptor in the (+) gamete was suggested by results with α-mannosidase which removes a terminal mannose required for gametic contact and probably also for concanavalin A binding. Conflicting results were obtained by McLean and Brown (1974) who also studied *C. moewusii* but found that mating of the (+) gamete is insensitive to (univalent) concanavalin A. The discrepancies in results between these two laboratories may be due to differences in the valence of the concanavalin A preparations used. The action of tetravalent lectin is to cross-link membrane glycoprotein components which would immobilize these and other membrane receptors. This immobilization might prevent aggregation of related molecules on the membrane surface required for flagellar tip agglutination. Monovalent lectins do not cross-link and therefore do not reduce mobility of membrane molecules (Gingell, 1973). Certainly such experimental problems need to be resolved before any meaningful conclusions can be reached concerning the nature of the recognition factors in *Chlamydomonas*. Ursula W. Goodenough will be reviewing 'Mating interactions in *Chlamydomonas*' for *Microbial Interactions*.

Differences in binding of Alcian Blue to mating types *5* and *21* of *Hansenula wingei* have been detected (G. Stewart, personal communication). Strain *5* bound 2.35 μg Alcian Blue mg^{-1} dry weight of yeast whereas strain *21* bound 2.02 μg mg^{-1}. Although these binding differences are statistically significant at the 95% level, they may not be biologically significant with respect to sex-specific agglutination properties because the *5* x *21* diploid synthesizes neither *5f* nor *21f* and yet it binds more dye (2.82 μg mg^{-1}) than either haploid. Therefore, it is likely that surface charge differences are due to other molecules unrelated to mating type. It is anticipated that Alcian Blue binding ability would segregate independent of mating type at meiosis. Basic (or cationic) dyes such as Alcian Blue and Congo Red bind better to strain *5* whereas acidic (or anionic) dyes such as Malachite Green bind better to strain *21* (M. Crandall, unpublished observations). It is anticipated, therefore, that these cell surface charge differences would be detectable by electrophoretic mobility of whole cells.

The two mating types of *Chlamydomonas moewusii* differ in electrophoretic mobility by −0.35 (μm s^{-1}) V^{-1} cm^{-1} (McLean and Bosmann, 1975). Unless the two mating types studied have been extensively backcrossed and are, therefore, isogenic, this difference in surface charge may be unrelated to mating type.

Differences in the cell surface charge of mating types can result in different binding affinities to resins as in the case of *Saccharomyces cerevisiae* strains **a** and α (Biliński and Litwińska, 1974). But the ability to separate opposite mating types by column chromatography may be related to physical rather than chemical differences, as in the case of bacteria. Donor and recipient *Escherichia coli* can be separated on mixed Cellex-P columns because the F pili cause the donor cells to become trapped in the cellulose whereas the females pass through the column; the separation pattern is unchanged by pH or salt concentration but if the males are pretreated in a blender to remove pili, then they too pass through the column (Zsigray *et al.*, 1970).

Antibodies synthesized against crude extracts or whole cells of gametes may not be directed toward the sex-specific receptor but toward another antigen fortuitously present in high amounts in one of the sexes. An example of one such antigen might be the protein identified as a band in electrophoresis of a preparation derived from the (+) agglutinin of *Chlamydomonas* (Snell, 1976b). This protein (70 000 daltons) is not found in (−) agglutinin (gamone) nor in similar extracts from vegetative cells. Since it is found in preparations of both mastigonemes and vesicular membranes (derived from gamone) its role in sex-specific adhesion is not known since mastigonemes are not involved in recognition. This protein might form part of the specialized mating structure (choanoid body) that is (i) associated with the cell membrane; (ii) involved in formation of the fertilization tube and (iii) found only in the (+) gamete (Goodenough and Weiss, 1975; Triemer and Brown Jr., 1975).

Antiserum prepared against one cell type can be adsorbed to the opposite cell type (or to the diploid) as a means of purifying for the mating type-specific antibody. This adsorption step will remove antibodies against antigens common to both cell types but may lead to problems. Mannan forms a considerable portion of both recognition factors from the yeast *Hansenula wingei* as well as of the yeast cell wall. If the sex-specific antibody were directed against the mannan moiety of the sex factor, then it might be lost during purification by adsorption.

In order to determine whether the antibodies or surface properties that differ between the two sexes are related to mating type specifically, the recognition factors must be extracted from the cell and the purification must be monitored using a biological assay such as agglutination or inhibition of agglutination (described in Crandall and Brock, 1968). While radiolabelling or immunological assays are valuable tools in the study of macromolecules because of their sensitivity, biological assays are essential in the identification of the recognition factors because of their specificity. Biological assays can also be quite sensitive; for example, cytoplasmic extracts containing 5-agglutinin are still active at a 1 : 1000 dilution and at concentrations below the detection limit of the Lowry protein assay (Crandall and Brock, 1968). Genetic analyses should accompany the biochemical studies of the recognition factors to determine whether the molecule in question is coded for or controlled by the mating type locus.

In the absence of a system amenable to genetic analysis or an isolation procedure that yields active receptor molecules, some information about the chemical nature of the recognition factors can be obtained from either the use of specific enzymes that destroy activity *in situ* or competitive inhibitors that are analogs of the combining site and prevent cell association. Information about the chemical nature of the recognition factors can also be obtained from the study of mutants lacking certain cell surface components. Ideally, conditional mutants such as temperature-sensitive mutants should be isolated to that the gain or loss of sexual agglutination or mating ability with shifts from restrictive to permissive conditions and vice versa can be correlated with concomitant changes in other surface components. Only through a combined approach involving

Table 2.9 Complementary receptors involved in cell recognition

Recognition factors	Organism	References
	Bacteria	
Carbohydrate portion of lipopolysaccharide plus outer membrane protein (female)	Escherichia coli	Lancaster et all., (1965);Schwarts et al., (1965);Monner et al., (1971); Wiedemann and Schmidt (1971); Skurray et al., (1974); Manning and Reeves, (1975; 1976); Henning et al., (1976); reviewed in Curtiss III et al., (1976); Lancaster, (personal communication)
Tip of F pilus (male)	Escherichia coli	Ou, (1975); reviewed in Curtiss III, et al., (1976)
	Algae	
Glycoproteins [both (+) and (−) mating types] located on flagellar membrane vesicles	Chlamydomonas spp.	Wiese and Hayward, (1972); McLean et al., (1974); Bergman et al., (1975); Snell, (1976b)
	Fungi	
Protein (high molecular weight on membrane of zygote after fusion of amoebae)	Didymium iridis	J. Yemma, (personal communication)
Mannanproteins (4×10^4 daltons from mating type 21 and 10^6 daltons from mating type 5)	Hansenula wingei	Crandall and Brock, (1968); Taylor and Orton, (1971); Yen and Ballou, (1973; 1974); Crandall et al., (1974); reviewed in Crandall et al., (1976)
Glycoproteins (both 10^6 daltons from both mating types a and α)	Saccharomyces cerevisiae	Shimoda et al.,(1975); Shimoda and Yanagishima, (1975)
Proteins [both (+) and (−)]	Schizosaccharomyces pombe	Calleja, (1974)

Table 2.9 Complementary receptors involved in cell recognition (*continued*)

Recognition factors	Organism	References
	Protozoa	
Protein receptor associated with lipid (10^6 daltons or larger from mating type I); glycoprotein (2×10^4 daltons from mating type II)	*Blepharisma intermedium*	Miyake and Honda, (1976)
Proteins (compatible mating types) located on ciliary membrane vesicles	*Paramecium* spp.	Clark, (1972); Kitamura and Hiwatashi, (1976)

physiological, biochemical, genetical, ultrastructural, immunological, enzymatic and mutational studies can any detailed information be gained in any one microbial mating system. Perhaps the time is ripe for researchers to make a concerted effort to study one or a few mating systems in depth. Those systems in which studies of the recognition factors have been initiated are presented in Table 2.9. The only microbial mating system in which both recognition factors have been purified and chemically characterized is the yeast *Hansenula wingei* (references cited in Table 2.9). Strains *5* and *21* of this yeast that agglutinate and mate with high efficiency have been backcrossed repeatedly to the wild type and testers have been selected that sporulate well in hybrids which yield ascospores that germinate with a high frequency (M. Crandall, unpublished) These testers are multiply-marked with auxotrophic and resistance mutations and are available for genetic and biochemical studies. Other microbial systems such as *Saccharomyces cerevisiae, Escherichia coli, Schizosaccharomyces pombe* and *Chamydomonas* have well developed genetic maps but biochemical studies of the mating type-specific receptors are lacking.

2.5 LOCALIZED ENZYMATIC CHANGES INDUCED BY CELL CONTACT

In all microbial mating systems, whether there is only transfer of genetic material from one sex to the other (as in bacteria and protozoa) or whether there is zygotic fusion of the two gametes (as in algae and fungi), there must be breakdown of the cell membrane (and wall, if present) at the point of cell contact. Very little information is available concerning the nature of the enzymes involved in this localized autolysis. Some release of carbohydrate and protein occurs during conjugation in yeast (reviewed in Crandall *et al.,* 1976) but it is not clear whether this is due to breakdown of cell wall material or leakage of cytoplasmic materials as a result of increased permeability of the cell membrane. Whatever enzymes are involved, there must be concomitant synthesis occurring with the autolysis or both cells would lyse.

Observations common to the process of cell fusion in different systems indicate a basic mechanism in the control of cell surface functions. Both cell fusion (in which two cells become one) and cell division (in which one cell becomes two) involve a highly regulated series of membrane alterations called surface modulations by Edelman, (1976). Such surface changes in membrane structure and function may be initiated by perisemic signals caused by pheromones or cell contact (discussed by Reissig, 1974). Edelman (1976) proposes that the processes of cell growth and recognition are co-ordinated by assembly at certain points on the membrane of inter-acting macromolecules consisting of mobile surface receptors and sub-membranous fibrillar structures (microtubules). An interesting observ-ation that may tie together this proposal by Edelman that surface modul-ation is governed by microtubules and the idea presented in Section 2.2.2 that calcium ion interferes with cell fusion is that calcium ions cause disassembly of microtubules (summarized in Marx, 1976). If microtubules are, in fact, transport vehicles that carry membrane and wall components from their cytoplasmic sites of synthesis (endoplasmic reticulum, Golgi apparatus) or storage (small vesicles) to the cell surface, then the study of effects of metal ions and pheromones on these fibrillar structures seems to be the next obvious direction to take in research in the enzym-ology of cell fusion.

2.6 BINDING THINGS TOGETHER (SUMMARY)

Sex is widespread in the invisible kingdom. Species with opposite mating type strains are distributed throughout each major group of the protists (bacteria, algae, fungi and protozoa). Conspicuously absent from this list are the blue-green algae in which only scattered reports of genetic recombination have appeared. Undoubtedly this lack of discovery of conjugation in blue-green algae is due to inherent difficulties in studying these organisms, many of which have never been cultured axenically.

Such extensive distribution of sexually-interacting systems among the other microbes signifies that the ability to mate favors the species for survival because it allows for genetic recombination. Mating efficiency is promoted by certain molecules such as complementary cell surface recognition factors, sex hormones (pheromones) and attractants and their corresponding receptors on the cell surface. Rather elaborate interacting chemical systems have been evolved by these 'primitive' microbial cells to facilitate the processes of: gametogenesis, attraction, recognition and fusion of opposite cell types. Similar interacting chemical systems known in the more 'advanced' mammalian cells may represent conservation of genetic information during evolution from simpler aggregating systems.

Commitment of microbial cells to either asexual or sexual reproduction is regulated by: pheromones, nutrient limitation, metal ions and exogenous chemicals, light, cytoplasmic as well as nuclear genes and the cell cycle. The chemical nature of pheromones varies as much as the morphological traits differ between the different taxonomic groups. Pheromones are secreted only by the eucaryotic protists and can be small molecules (steroids or other lipids, organic acid derivatives or peptides) or large molecules (glycoproteins). Steroid and peptide pheromones are found only among the fungi. These same types of compounds function as hormones in mammalian systems also.

Microbial cells that do not secrete pheromones are dependent on the proper environmental conditions for the induction of sexual differentiation. Nitrogen starvation appears to be the single most important stimuli for mating, with a low calcium content of the medium acting, in concert to induce cell fusion. Limitation of nutrients or the excretion of pheromones function to synchronize sexual partners at Gl of the cell division cycle. At this point, each cell has a complete haploid chromosomal set and subsequent cell fusion will result in a genetically balanced diploid organism.

Once the vegetative cell has differentiated, the encounter between gametes can be fortuitous or can be promoted by chemical attractants. All the chemotropic or chemotactic agents that have been characterized are lipids; some cause non-motile cells of opposite sex to bend or grow toward one another whereas others cause motile gametes to swim toward the opposite sex. In many cases, the same pheromone that induced the vegetative cell to undergo gametogenesis, then acts as a chemotactic agent for the gametes.

The first parts of each gamete to touch are filamentous appendages that cause opposite sexes to bind together. Specificity of cell contact is governed by glycoprotein recognition factors located on these surface filaments. Recognition factors are complementary and form a neutralize complex that bridges the two apposed cell surfaces. Similar cell recogniti events occur in many different tissues of an animal body (see Receptors and Recognition, Vol. 1., Greaves).

Following the initial cell recognition step, the two microbial gametes become joined more intimately by a conjugation bridge that forms by fusion of the outer surface layers of both cells. The enzymatic changes leading to this tight cell binding and ultimate fusion are highly localized and involve both degradative and synthetic reactions. It is thought that these reactions involve: aggregation of mobile membrane components; changes in membrane permeability; contributions of structural glycoproteins from Golgi vesicles by exocytosis and intracellular communication via microtubules. However, all the functional enzymes and structural components involved in this process remain to be identified.

Future research in the area of mating type interactions in micro-organisms will undoubtedly concentrate on the identification of the chemical and structural components and enzymes involved in each step in mating. If, at this time, scientists join research efforts and collaborate on a few microbial systems, in depth progress toward the understanding of cell interactions could, conceivably, be attained in the foreseeable future.

REFERENCES

Abe, K., Kusaka, I. and Fukui, S. (1975), *J. Bact.,* **122,** 710–718.

Al-Hassan, K.K. and Fergus, C.L. (1969), *Mycopath. Mycol. appl.,* **39,** 273–286.

Bandoni, R.J. (1965), *Can. J. Bot.,* **43,** 627–630.

Barksdale, A.W. (1962), *Am. J. Bot.,* **49,** 633–638.

Barksdale, A.W. (1969), *Science,* **166,** 831–837.

Barksdale, A.W. and Lasure, L.L. (1974), *Appl. Microbial,* **28,** 544–546.

Bergman, K., Burke, P.V., Cerdá-Olmedo, E., Davic C.N., Delbrück, M., Foster, K.W., Goodell, E.W., Heisenberg, M., Meissner, G., Zalokar, M., Dennison, D.S. and Shropshire, Jr., W. (1969), *Bact. Rev.,* **33,** 99–157.

Bergman, K., Goodenough, U.W., Goodenough, D.A., Jawitz, J. and Martin, H. (1975), *J. Cell Biol.,* **67,** 606–622.

Biliński, T. and Litwińska, J. (1974), *Bull. Acad. Pol. Sci. Cl. VI Ser. Sci. biol.* **22,** 173–176.

Biliński, T., Litwińska, J., Żuk, J. and Gajewski, W. (1975), Synchronous zygote formation in yeasts. In: *Methods in Cell Biology,* D.M. Prescott, ed. Vol. 11, pp. 89–96, Academic Press, New York.

Bishop, H. (1940), *Mycologia,* **32,** 505–529.

Bistis, G.N. and Raper, J.R. (1963), *Am. J. Bot.,* **50,** 880–891.

Bradley, D.E. (1972), *Genet. Res.,* **19,** 39–51.

Braun, V. and Miyake, A. (1975), *F.E.B.S. Letters,* **53,** 131–134.

Brinton, C.C. (1965), *Trans. N.Y. Acad. Sci.,* **27,** 1003–1054.

Brock, T.D. (1965), *Proc. natn. Acad. Sci., U.S.A.,* **54,** 1104–1112.

Brown, Jr., R.M., Johnson, C. and Bold, H.C. (1968), *J. Phycol.,* **4,** 100–120.

Bu'Lock, J.D. (1967), *Essays in Biosynthesis and Microbial Development,* John Wiley and Sons. New York.

Bu'Lock, J.D. (1976), Hormones in fungi. In: *The Filamentous Fungi,* J.E. Smith and D.R. Berry, eds. Edward Arnold, London.

Bu'Lock, J.D., Jones, B.E., Quarrie, S.A. and Winstanley, D.J. (1974), *Arch. Microbiol.,* **97,** 239–244.

Bu'Lock, J.D., Jones, B.E. and Winskill, N. (1976), *Pure Appl. Chem.*, **37**, in press.

Calleja, G.B. (1973), *Archs. Biochem. Biophys.*, **154**, 382–386.

Calleja, G.B. (1974), *Can. J. Microbiol.*, **20**, 797–803.

Calleja, G.B., Johnson, B.F. and Yoo, B.Y. (1976), *J. Cell Biol.*, **70**, 132a (abstract).

Cantino, E.C. (1966), Morphogenesis in aquatic fungi. In: *The Fungi, An Advanced Treatise, The Fungal Organism,* G.C. Ainsworth and A.S. Sussman, eds. Vol. 2, pp. 283–337, Academic Press, New York.

Child, J.J. and Haskins, R.H. (1971), *Can. J. Bot.*, **49**, 329–332.

Clark, M.A. (1972), *J. Cell Physiol.*, **79**, 1–14.

Clark, M.A., Francis, D. and Eisenberg, R. (1973), *Biochem. biophys. Res. Commun.*, **52**, 672–678.

Cohen, L.W. (1965), *Expl. Cell Res.*, **37**, 360–367.

Comfort, A. (1972), *The Joy of Sex,* Crown Publishing, New York.

Cook, W.J. and Bugg, C.E. (1975), *Biochim. biophys. Acta.* **389**, 428–435.

Crandall, M.A. and Brock, T.D. (1968), *Bact. Rev.*, **32**, 139–163.

Crandall, M., Lawrence, L.M. and Saunders, R.M. (1974), *Proc. natn. Acad. Sci., U.S.A.*, **71**, 26–29.

Crandall, M. and Caulton, J.H. (1975), Induction of haploid glycoprotein mating factors in diploid yeast. In: *Methods in Cell Biology,* D.M. Prescott, ed. Vol. 12, pp. 185–207. Academic Press, New York.

Crandall, M., Egel, R. and MacKay, V.L. (1976), Physiology of mating in three yeasts. In: *Advances in Microbial Physiology,* A.H. Rose and D.W. Tempest, eds., Academic Press, London, (in press).

Cronkite, D.L. (1976), *J. Protozool.* **23**, 431–432.

Cummins, J.E. and Day, A.W. (1973), *Nature,* **245**, 259–260.

Cummins, J.E. and Day, A.W. (1974), The cell cycle regulation of sexual morphogenesis in a basidiomycete *Ustilago violaces.* In: *Cell Cycle Controls,* G.M. Padilla, I.L. Cameron, and A. Zimmerman, eds., pp. 181–200. Academic Press, New York.

Curtiss, III, R. (1969), *Ann. Rev. Microbiol.*, **23**, 69–136.

Curtiss, III, R., Cars, L.G., Allison, D.P. and Stallions, D.R. (1969), *J. Bact.*, **100**, 1091–1104.

Curtiss, III, R., Fenwick, Jr., R.G., Goldschmidt, R. and Falkinham, III, J.O. (1976), The mechanism of conjugation. In: *Transferable Drug Resistance Factor R.* ed. S. Mitsuhashi, University Park Press, Baltimore, (in press).

Darden, W.H. (1970), *Ann. N.Y. Acad. Sci.*, **175**, 757–763.

Darden, W.H. (1973a), Hormonal control of sexuality in algae, In: *Humoral Control of Growth and Differentiation,* J. Lobue, and S. Gordon, eds., Vol. 2, pp. 101–119. Academic Press, New York.

Darden, W.H. (1973b), *Microbios.*, **8**, 167–174.

Day, A.W. (1972), *Nature New Biol.*, **237**, 282–283.

Day, A.W. and Poon, N.H. (1975), *Can. J. Microbiol.*, **21**, 547–557.

Doerder, F.P. and Debault, L.E. (1975), *J. Cell Sci.*, **17**, 471–493.

Dring, M.J. (1974), Reproduction In: *Algal Physiology and Biochemistry*.
W.D.P. Stewart, ed. pp. 814–837, University of California Press, Berkeley.
Driver, C.H. and Wheeler, H.E. (1955), *Mycologia*, **47**, 311–316.
Duntze, W. (1974), *Postepy Mikrobiologiya, Tom XIII Ziszyt* **2**, 41–51.
Duntze, W., MacKay, V. and Manney, T.R. (1970), *Science*, **168**, 1472–1473.
Duntze, W., Stotzler, D., Bücking-Throm, E. and Kalbitzer, S. (1973), *Eur. J. Biochem.*,
35, 357–365.
Edelman, G.M. (1976), *Science*, **192**, 218–226.
Elliott, C.G. (1972), *J. Gen. Microbiol.*, **72**, 321–327.
Ellis, R.J. and Machlis, L. (1968), *Am. J. Bot.*, **55**, 600–610.
Erdos, G.W., Raper, K.B. and Vogen, L.K. (1973), *Proc. natn. Acad. Sci. U.S.A.*,
70, 1828–1830.
Esser, K. (1966), Incompatibility. In: *The Fungi, An Advanced Treatise. The
Fungal Organism*, G.C. Ainsworth and A.S. Sussman, eds., Vol. 2,
pp. 661–676, Academic Press, New York.
Filosa, M.F., Kent, S.G. and Gilette, M.U. (1975), *Devl. Biol.*, **46**, 49–55.
Flegel, T.W. (1976), *Can. J. Bot.*, **54**, 411–418.
Fowell, R.R. (1969a), Sporulation and hybridization of yeasts. In: *The Yeasts,
Biology of Yeasts*, A.H. Rose and J.S. Harrison, eds., Vol. 1, pp. 303–383.
Academic Press, London.
Fowell, R.R. (1969b), Life cycles in yeasts. In: *The Yeasts, Biology of Yeasts*,
A.H. Rose and J.S. Harrison, eds., Vol. 1, pp. 461–471. Academic Press,
London.
Giese, A.C. (1973), *Blepharisma, the biology of a light sensitive protozoan*.
Stanford University Press, Stanford.
Gingell, D. (1973), *J. theor. Biol.*, **38**, 677–679.
Gooday, G.W. (1974), *Ann. Rev. Biochem.*, **43**, 35–50.
Gooday, G.W. (1975), Chemotaxis and chemotropism in fungi and algae. In: *Primitive
Sensory and Communication Systems*, M.J. Carlie, ed., pp. 155–204.
Academic Press, London.
Goodenough, U.W. and Weiss, R.L. (1975), *J. Cell Biol.*, **67**, 623–637.
Harris, R.H. and Mitchell, R. (1973), *Ann. Rev. Microbiol.*, **27**, 27–50.
Hawker, L.E. (1966), Environmental influences on reproduction. In: *The Fungi,
The Fungal Organism*, G.C. Ainsworth and A.S. Sussman, eds., Vol. 2,
pp. 435–469, Academic Press, New York.
Hendrix, J.W. (1970), *Ann. Rev. Phytopath.*, **8**, 111–130.
Henning, U., Hindennach, I. and Haller, I. (1976), *F.E.B.S. Letters*, **61**, 46–48.
Herman, A.I. (1971a), Antonie van Leeuwenhoek, *J. Microbiol. Serol.*, **37**,
379–384.
Herman, A.I. (1971b), *J. Bact.*, **107**, 371.
Hicks, J.B. and Herskowitz, I. (1976), *Nature*, **260**, 246–248.
Hill, G.J.C. and Machlis, L. (1970), *Pl. Physiol.*, **46**, 224–226.
Hiwatashi, K. (1969), *Paramecium*. In: *Fertilization, Comparative Morphology,
Biochemistry and Immunology*, C.B. Mets and A. Monroy, eds., Vol. 2,
pp. 255–294, Academic Press, New York.

Hlubucek, J.R., Hora, J., Toube, T.P. and Weedon, B.C.L. (1970), *Tetrahedron Lett.*, **59**, 5163–5164.
Honda, H. and Miyake, A. (1975), **257**, 678–680.
Honour, R.C. and Tsao, P.H. (1974), *Mycologia*, **66**, 1030–1038.
Hopwood, D.A., Chater, K.F., Dowding, J.E. and Vivian, A. (1972), *Bact. Rev.*, **37**, 371–405.
Horenstein, E.A. and Cantino, E.C. (1969), Fungi. In: *Fertilization, Comparative Morphology, Biochemistry and Immunology*, C.B. Metz and A. Monroy, eds., Vol. 2, pp. 95–133. Academic Press, New York.
Horgen, P.A. (1976), Steroid induction of differentiation: *Achlya* as a model system. In: *Eukaryotic Microbes as Model Development Systems*, D.H. O'Day and P.A. Horgen, eds., Marcel Dekker, New York, (in press).
Horgen, P.A. and Ball, S.F. (1974), *Cytobios*, **10**, 181–186.
Horowitz, D.K. and Russell, P.J. (1974), *Can. J. Microbiol.*, **20**, 977–980.
Imafuku, M. (1975), *J. Interdiscipl. Cycle Res.*, **6**, 141–152.
Jaenicke, L. and Müller, D.G. (1973), *Fortschr. Chem. Org. Naturst.*, **30**, 61–100.
Jaenicke, L., Müller, D.G. and Moore, R.E. (1974), *J. Am. Chem. Soc.*, **96**, 3324–3325.
Jaenicke, L. and Seferiadis, K. (1975), *Chem. Ber.*, **108**, 225–232.
Jones, R.F., Kates, J.R. and Keller, S.J. (1968), *Biochim. biophys. Acta*, **157**, 589–598.
Karlson, P. and Luscher, M. (1959), *Nature*, **183**, 55–56.
Kates, J.R. and Jones, R.F. (1964), *J. Cell comp. Physiol.*, **63**, 157–164.
Kimball, R.F. (1939), *Am. Nat.*, **73**, 57–71.
Kitamura, A. and Hiwatashi, K. (1976), *J. Cell Biol.*, **69**, 736–740.
Kochert, G. (1975), Developmental mechanisms in *Volvox* reproduction. In: *The Developmental Biology of Reproduction*, C.L. Markert and J. Papaconstantinous, eds., pp. 55–90, Academic Press, New York.
Kochert, G. and Yates, I. (1974), *Proc. natn. Acad. Sci., U.S.A.*, **71**, 1211–1214.
Kubota, T., Tokoroyama, T., Tsukuda, Y., Koyama, H. and Miyake, A. (1973), *Science*, **179**, 400–402.
Lancaster, J.H. Goldschmidt, E.P. and Wyss, O. (1965), *J. Bact.*, **89**, 1478–1481.
Levi, J.D. (1956), *Nature*, **177**, 753–754.
Lewis, K.E. and O'Day, D.H. (1976), *Can. J. Microbiol.*, (in press).
Loewenstein, W.R. (1972), Cell-to-cell connections. In: *Cell Interactions*, L.G. Silvestri, ed., pp. 296–298. North-Holland, Amsterdam.
Luporini, P. and Dini, F. (1975), *J. Protozool*, **22**, 541–544.
MacHac, M.A. and Bonner, J.T. (1975), *J. Bact.*, **124**, 1624–1625.
Machlis, L. (1966), Sex hormones in fungi. In: *The Fungi, An Advanced Treatise, The Fungal Organism*, G.C. Ainsworth and A.S. Sussman, eds., Vol. 2, pp. 415–431. Academic Press, New York.
MacKay, V. and Manney, T.R. (1974), *Genetics*, **76**, 255–271.
Manning, P.A. and Reeves, P. (1975), *J. Bact.*, **124**, 576–577.

Manning, P.A. and Reeves, P. (1976), *J. Bact.*, **127**, 1070–1079.

Marvin, D.A. and Hohn, B. (1969), *Bact. Rev.*, **33**, 172–209.

Marx, J.L. (1976), *Science*, **192**, 455–457.

Mas, J., Celis, E., Piña, E. and Brunner, A. (1974), *Biochem. biophys. Res. Commun.*, **61**, 613–620.

Mayer, F. (1971), *Arch. Mikrobiol.*, **76**, 166–173.

McLean, R.J.and Bosmann, H.B. (1975), *Proc. natn. Acad. Sci. U.S.A.*, **72**, 310–313.

McLean, R.J.and Brown, Jr., R.M. (1974), *Devl. Biol.*, **36**, 279–285.

McLean, R.J., Laurendi, C.J. and Brown, Jr., R.M. (1974), *Proc. natn. Acad. Sci.*, *U.S.A.*, **71**, 2610–2613.

McMorris, T.C., Seshadri, R., Weihe, G.R., Arsenault, G.P. and Barksdale, A.W. (1975), *J. Am. chem. Soc.*, **97**, 2544–2545.

Mesland, D.A.M., Huisman, J.G. and van cen Ende, H. (1974), *J. gen. Microbiol.*, **80**, 111–117.

Metz, C.B. (1954), Mating substances and the physiology of fertilization in ciliates. In: *Sex in Micro-organisms*, D.H. Wenrich, ed., pp. 284–334, A.A.A.S. Washington, D.C.

Meynell, E., Matthews, R.A. and Lawn, A.M. (1974), *J. gen. Microbiol.*, **82**, 203–205.

Miyake, A. (1964), *Science*, **146**, 1583–1585.

Miyake, A. (1974), *Curr. Topics Microbiol. Immunol.*, **64**, 49–77.

Miyake, A. (1975), *Science*, **189**, 53–55.

Miyake, A. and Beyer, J. (1973), *Expl. Cell Res.*, **76**, 15–24.

Miyake, A. and Bleyman, L.K. (1976), *Genet. Res.*, **27**, 267–276.

Miyake, A. and Honda, H. (1976), *Expl. Cell Res.*, (in press).

Miyake, A. and Nobili, R. (1974), *J. Protozool.*, **21**, 584–587.

Monner, D.A., Jonsson, S. and Boman, H.G. (1971), *J. Bact.*, **107**, 420–432.

Morris, E.O. (1966), Aggregation of unicells: yeasts. In: *The Fungi, An Advanced Treatise, The Fungal Organism*, E.C. Ainsworth and A.S. Sussman, eds., Vol. 2, pp. 63–82. Academic Press, New York.

Müller, D.G. (1972), *Ber. D. bot. Ges.*, **85**, 363–369.

Müller, D.G. (1974), *Biochem. Physiol. Pflanz*, **165**, 212–215.

Müller, D.G., Jaenicke, L., Donike, M. and Akintobi, T. (1971), *Science*, **171**, 815–817.

Mullins, J.T. and Warren, C.O. (1975), *Am. J. Bot.*, **62**, 770–774.

Nutting, W.H., Rapoport, H. and Machlis, L. (1968), *J. Am. chem. Soc.*, **90**, 6434–6438.

O'Day, D.H. and Lewis, K.E. (1975), *Nature*, **254**, 431–432.

Okazaki, K. and Holtzer, H. (1966), *Proc. natn. Acad. Sci. U.S.A.*, **56**, 1484–1490.

O'Malley, B.W. and Means, A.R. (1974), *Science*, **183**, 610–620.

Osumi, M., Shimoda, C. and Yanagishima, N. (1974), *Arch. Microbiol.*, **97**, 27–38.

Ou, J.T. (1973), *J. Bact.*, **115**, 648–654.

Ou, J.T. (1975), *Proc. natn. Acad. Sci. U.S.A.*, **72**, 3721–3725.

Ou, J.T. and Andersons, T.F. (1970),*J. Bact.,* **102**, 648–654.

Ou, J.T. and Reim, R. (1976), *J. Bact.,* **128**, (in press).

Ouyezdsky, K.B. and Szanislo, P.J. (1973),*J. Bact.,* **114**, 1356–1358.

Padan, E. and Shilo, M. (1973), *Bact. Rev.,* **37**, 343–370.

Pall, L. (1974), *Biochem. biophys. Res. Commun.,* **57**, 683–688.

Pauling, L. (1974), *Nature,* **248**, 769–771.

Phillips, R.B. (1971),*J. Protozool,* **18**, 163–165.

Poon, N.H., Martin, J. and Day, A.W. (1974), *Can. J. Microbiol.,* **20**, 187–191.

Preer, Jr., J.R. (1971), Genetics of the protozoa. In: *Research In Protozoology,*
    Tze-Tuan Chen, ed., Vol. 3, pp. 129–278. Pergamon Press, Oxford.

Rayburn, W.R. (1974),*J. Phycol.,* **10**, 258–265.

Raper, J.R. (1966), Life cycles, basic patterns of sexuality, and sexual mechanisms.
    In: *The Fungi, An Advanced Treatise, The Fungal Organism,* E.C. Ainsworth
    and A.S. Sussman, eds., Vol. 2, pp. 473–511. Academic Press, New York.

Reid, I.D. (1974), *Can. J. Bot.,* **52**, 521–524.

Reissig, J.L. (1974), Decoding of regulatory signals at the microbial surface.
    In: *Current Topics in Microbiology and Immunology,* W. Arber *et al.,* eds.,
    Vol. 67, pp. 43–96. Springer-Verlag, New York.

Revel, J.-P. and Ito, S. (1967), The surface components of cells. In: *The Specificity
    of Cell Surfaces,* B.D. Davis and L. Warren, eds., pp. 211–234. Prentice-
    Hall, Englewood Cliffs, N.J.

Ribeiro, O.K., Erwin, D.C. and Zentmyer, G.A. (1975),*Mycologia,* **67**, 1012–1019.

Ricci, N., Esposito, F. and Nobili, R. (1975),*J. exp. Zool.,* **192**, 343–348.

Ricci, N., Esposito, F., Nobili, R. and Revoltella, R. (1976),*J. Cell Physiol.,*
    87, 363–370.

Roth, S., McGuire, E.J. and Roseman, S. (1971),*J. Cell Biol.,* **51**, 536–547.

Salvin, S.B. (1942), *Am. J. Bot.,* **29**, 97–104.

Schmeisser, E.T., Baumgarter, D.M. and Howell, S.H. (1973), *Devl. Biol.,* **31**, 31–37.

Schwartz, G.H., Eiler, D. and Kern, M. (1965),*J. Bact.,* **89**, 89–94.

Sena, E.P., Radin, D.N., Welch, J. and Fogel, S. (1975), Synchronous mating in
    yeasts. In: *Methods in Cell Biology,* D.M. Prescott, ed., Vol. 11, pp. 71–88.
    Academic Press, New York.

Sermonti, G. (1969), Bacteria. In: *Fertilization, Comparative Morphology,
    Biochemistry and Immunology,* C.B. Metz and A. Monroy, eds., Vol. 2,
    pp. 47–94. Academic Press, New York.

Sherwood, W.A. (1966),*Mycologia,* **58**, 215–220.

Shimoda, C., Kitano, S. and Yanagishima, N. (1975), *Antonie van Leeuwenhoek
    J. Microbiol. Serol.,* **41**, 513–519.

Shimoda, C. and Yanagishima, N. (1975), *Antonie van Leeuwenhoek, J. Microbiol.
    Serol.,* **41**, 521–532.

Sing, V., Yeh, Y.-F. and Ballou, C.E. (1976), Isolation of a *Hansenula wingei*
    mutant with an altered sexual agglutinin, paper presented at the NATO
    Advanced Study Institute on Surface Membrane Receptors: Interface
    between cells and their environment, Bellagio, Italy, September 12–21,
    1975.

Skurray, R.A. and Reeves, P. (1973), *J. Bact.*, **114**, 11–17.

Skurray, R.A., Hancock, R.E.W. and Reeves, P. (1974), *J. Bact.*, **119**, 726–735.

Snell, W.J. (1976a), *J. Cell Biol.*, **68**, 70–79.

Snell, W.J. (1976b), *J. Cell Biol.*, **68**, 48–69.

Sonneborn, T.M. (1963), *J. Protozool.*, **10** (suppl), 25.

Sonneborn, T.M. (1974a), *Paramecium aurelia.* In: *Handbook of Genetics Plants, Plant Viruses, and Protists,* R.C. King, ed., Vol. 2, pp. 469–594, Plenum Press, New York.

Sonneborn, T.M. (1974b), *Tetrahymena pyriformis.* In: *Handbook of Genetics, Plants, Plant Viruses, and Protists,* R.C. King, ed., Vol. 2, pp. 433–467. Plenum Press, New York.

Starr, R.C. and Jaenicke, L. (1974), *Proc. natn. Acad. Sci.*, **71**, 1050–1054.

Steele, R.L. (1965), *Bioscience,* **15**, 298.

Steinhardt, R.A. and Epel, D. (1974), *Proc. natn. Acad. Sci.*, **71**, 1915–1919.

Streiblova, E. and Wolf, A. (1975), *Can. J. Microbiol.*, **21**, 1399–1405.

Sutter, R.P. (1975), *Proc. natn. Acad. Sci. U.S.A.*, **72**, 127–130.

Sutter, R.P. (1976), Regulation of the first stage of sexual development in *Phycomyces blakesleeanus* and in other mucoraceous fungi. In: *Eukaryotic Microbes as Model Developmental Systems,* D.H. O'Day and P.A. Horgen, eds., Marcel Dekker, New York, (in press).

Sutter, R.P., Capage, D.A., Harrison, T.L. and Keen, W.A. (1973), *J. Bact.*, **114**, 1074–1082.

Takahashi, M. and Hiwatashi, K. (1974), *Expl. Cell Res.*, **85**, 23–30.

Takahashi, M., Takeuchi, N. and Hiwatashi, K. (1974), *Expl. Cell Res.*, **87**, 415–416.

Taylor, N.W. and Orton, W.L. (1968), *Archs. Biochem. Biophys.*, **126**, 912–921.

Taylor, N.W. and Orton, W.L. (1970), *Biochem.* **9**, 2931–2934.

Taylor, N.W. and Orton, W.L. (1971), *Biochem.*, **10**, 2043–2049.

Taylor, N.W. and Tobin, R. (1966), *Archs. Biochem. Biophys.*, **115**, 271–276.

Trainor, F.R. and Burg, C.A. (1965), *Science,* **148**, 1094–1095.

Triemer, R.E. and Brown, Jr., R.M. (1975), *Protoplasma,* **85**, 99–108.

Tsubo, Y. (1961), *J. Protozool,* **8**, 114–121.

Turian, G. (1966), Morphogenesis in Ascomycetes. In: *The Fungi, An Advanced Treatise, The Fungal Organism,* G.C. Ainsworth and A.S. Sussman, eds., Vol. 2, pp. 339–385. Academic Press, New York.

van den Ende, H. (1976), Sexual morphogenesis in the *Phycomycetes,* In: *The Filamentous Fungi,* J.E. Smith and D.R. Berry, eds., Vol. 3, Edward Arnold, London, (in press).

van den Ende, H., Werkman, B.A. and Van den Briel, M.L. (1972), *Ark. Mikrobiol.,* **86**, 175–184.

Vaughan, V. and Barnett, A. (1973), *J. Protozool.,* **20**, 521–522.

Vonderhaar, B.K. and Topper, Y.J. (1974), *J. Cell Biol.,* **63**, 707–712.

Wardi, A.H. and Allen, W.S. (1972), *Analyt. Biochem.,* **48**, 621–623.

Wiedemann, B. and Schmidt, G. (1971), Structure and recipient ability in *Escherichia coli* mutants *Ann. N.Y. Acad. Sci.,* **182**, 123–125.

Wiese, L. (1969), Algae. In: *Fertilization, Comparative Morphology, Biochemistry and Immunology,* C.B. Metz and A. Monroy, eds., Vol. 2, pp. 135–188. Academic Press, New York.

Wiese, L. (1974), *Ann. N.Y. Acad. Sci.,* **234,** 383–395.

Wiese, L. and Hayward, P.C. (1972), *Am. J. Bot.,* **59,** 530–536.

Wiese, L. and Jones, R.F. (1963), *J. Cell Physiol.,* **61,** 265–274.

Wiese, L. and Shoemaker, D.W. (1970), *Biol. Bull.* **138,** 88–95.

Wilkinson, L.E. and Pringle, J.R. (1974), *Expl. Cell Res.,* **89,** 175–187.

Wolfe, J. (1974), *Expl. Cell Res.,* **87,** 39–46.

Wolfe, J. (1976), *Devl. Biol.,* (in press).

Wolk, C.P. (1973), *Bact. Rev.,* **37,** 32–101.

Wurtz, T. and Jockusch, H. (1975), *Devl. Biol.,* **43,** 213–220.

Yanagishima, N. (1971), *Physiol. Plant.,* **24,** 260–263.

Yen, P.H. and Ballou, C.E. (1973), *J. biol. Chem.,* **248,** 8316–8318.

Yen, P.H. and Ballou, C.B. (1974), *Biochem.,* **13,** 2428–2437.

Yoo, B.Y., Calleja, G.B. and Johnson, B.F. (1971), *Proc. Can. Fed. Biol. Soc.,* **14,** 62 (abstract).

Zickler, H. (1952), *Arch. Protistenk,* **98,** 1–70.

Zonneveld, B.J.M. (1975), *Arch. Microbiol.,* **105,** 101–104.

Zsigray, R.M., Fulk, G.E. and Lawton, W.D. (1970), *J. Bact.,* **103,** 302–304.

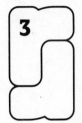

Erythrocyte proteins

HEINZ FURTHMAYR
Department of Pathology,
Yale University School of Medicine
New Haven, Connecticut USA 06510

Receptors and Recognition, Series A, Volume 3
Edited by P. Cuatrecasas and M.F. Greaves
Published in 1977 by Chapman and Hall, 11 New Fetter Lane, London EC4P 4EE
© 1977 Chapman and Hall

3.1 INTRODUCTION

The central dogma in molecular biology states that the flow of information from the fundamental deposit held in the DNA code of each cell nucleus directs the process of protein synthesis: once 'information' has passed into protein, it cannot get out again. Forbidden is the passage of genetic information from protein to protein or protein to nucleic acid. There are considerable arguments in the field of differentiation and morphogenesis focusing on the point of whether and to what extent these processes are initiated and determined by selective gene action or by epigenetic events [1]. Information as an epigenetic event can be received and translated by cells into expression of genes during ontogeny and this may be necessary for maintaining a fixed state of differentiation once it is achieved. There is much evidence indicating that differentiated cells which are normally fixed in organs or tissues tend to lose morphological individuality and specialized patterns of macromolecular synthesis and viability, if they are not supported in their differentiated state by their normal environment. The working assumption is that the cell membranes contain specific receptors which serve to accept stimuli and to transfer information, derived from the interaction between the receptor and its ligand, by unknown mechanisms across the membrane to produce specific changes in the state of the cell.

Many efforts have been made and are currently being pursued which try to establish links between properties of cells, at different stages in development, and specific macromolecules expressed on the surface of the plasma membrane. The presence of such macromolecules, their number, arrangement, or distribution may provide the basis for interaction between cells and their cellular or acellular environment in a flexible and changing way which eventually results in definitive spacial relationships.

These ideas imply that the composition of the cell membrane or the cell surface pattern is subject to variations due to stage of differentiation or environmental factors. This imposes severe restrictions with regard to the chemical analysis of the molecules involved. 'Information' can be stored in the membrane possibly in many ways and we have no means as yet to distinguish between them: (a) new proteins or glycoproteins are inserted into the membrane at various stages of development; (b) the

103

number or distribution of a given molecule changes; (c) post-translational information is added by specific enzymes which add or remove carbohydrates from glycoproteins or glycolipids; (d) proteins are secreted but are re-adsorbed onto the membrane.

The choice of the red cell membrane is a matter of convenience, which has allowed us to gain some important insight into the chemistry and molecular architecture of the mammalian plasma membrane, without having to pay too much attention to the complexities mentioned above. There are distinct advantages in working with highly differentiated, specialized membrane systems such as the red cell membrane, the photoreceptor membrane in retinal cells and halophilic bacteria, or the membrane of enveloped viruses in so far as only a small number of different proteins is associated with them. The limitations are that these systems can only provide partial answers to our understanding of the complex functional role of plasma membranes. The red cell in adult mammalian species is an end cell, the terminal differentiation of which is characterized by nuclear expulsion prior to the loss of cytoplasmic ribosomes and mitochondria [2]. Since little or no active biosynthesis occurs, all the equipment required for the functional period of about 120 days has been stored in the cell and the cell membrane. It is not known what elements the plasma membrane loses or acquires in the terminal step of differentiation. The assumption is made, however, that a population of cells can be collected which is rather uniform with regard to the content of major membrane proteins. The red cell membrane at the present time probably represents the best and most completely studied membrane regarding its composition of protein components and their relationship to each other. The surface characteristics of the erythrocyte have been studied extensively. From the time of Landsteiner's demonstration of the ABO blood groups as early as 1900, transplantation genetics and serology has abounded with examples of highly polymorphic and antigenic cell surface components. But only for a few systems is the chemical nature of these surface antigens known.

Most of our information has been obtained for the red cell from humans and to a large extent this article will be restricted to this system. Most major proteins in erythrocyte membranes of other mammalian species appear to be very similar to the proteins in humans. The notable exception is found within a class of heavily glycosylated proteins, which are expressed at the cell surface.

3.2 MAJOR PROTEINS IN THE HUMAN
RED CELL MEMBRANE

The last few years have seen a considerable expansion in our knowledge of membrane structure and membrane proteins. While not too long ago not a single homogeneous membrane protein was available for study in isolated form, this situation has changed dramatically. In several membrane systems, proteins have been purified to variable extents and structural analyses have been made. This development has changed our views on membrane structure considerably and old ideas on membrane structure were replaced with a more recent model, more consistent with the new information [3]. More or less empirically, it was found that some membrane proteins can be differentially released from the plasma membrane and other biomembranes by rather simple and mild procedures, while other proteins are released and solubilized only with agents which disrupt or destroy the lipid bilayer [4]. The latter class of proteins has been termed 'intrinsic or integral', and these terms were meant to distinguish proteins which are associated with the hydrophobic core of the lipid bilayer, from 'external or extrinsic' proteins, which are more 'loosely' bound to the exterior faces of the membrane via electrostatic interaction or other forces. Although rather convenient, these operational terms cannot adequately describe the mode of association with the bilayer lipid and eventually will have to be replaced with a more precise description in terms of lipid-protein, or protein-protein interaction, by hydrophobic, electrostatic, or other forces in relation to the bilayer.

The second part of this model states, that protein components within the lipid bilayer are mobile and can diffuse randomly and independently from each other within the plane of the membrane. The ratio of diffusion is largely determined by the nature and composition of the lipids in the bilayer. This implies that there is little if any ordered arrangement of proteins at the surface of the plasma membrane. There are however examples of highly ordered membrane structures, like the purple membrane in halobacteria, synapses or tight junctions. These may be clearly special cases of differentiated structures, for which order is introduced by either specific interactions between proteins within or outside the lipid bilayer or by interactions with submembranous structures. The absence of such highly ordered structures in most biological membranes does not preclude reversible interactions or other forces, the function of which would be to maintain a certain topographical relationship between various proteins. Thus far, such interactions though postulated have not been defined in any membrane system. On the other hand, lateral diffusion within the plane of the membrane has been observed using macromolecular ligands, for a number of cell surface markers, e.g., IgG on lymphocytes, 'lectin receptors', but in many instances the nature of the molecule and/or the mode of the interaction with the lipid bilayer have not been defined [68]. It remains to be seen whether these observations in the presently rather crude experiments will find support by more definitive studies.

The red cell membrane provides one model membrane system on which these ideas have been and can be further explored. The human ghost membranes contain about eight major polypeptide chains, when analyzed by sodium dodecyl sulfate gel electrophoresis (Fig. 3.1), some of which have been isolated and studied in greater detail. For most of the major polypeptides the biological function has still not been rigorously established and most commonly the numerical designation used by Fairbanks *et al.,* [5] has been adopted. The use of SDS gel electrophoresis as a rather simple analytical tool was instrumental for the development of the major concepts, despite a certain unreliability in determining the correct molecular weights. This became aparent when additional studies were done on certain membrane proteins.

Several experimental approaches have been developed to study the location of these major polypeptides with respect to the exterior or interior side of the membrane, and their relationship to each other. These include the treatment of intact cells or 'leaky' membranes with proteases, mono- or bifunctional chemical probes or with enzymes or enzyme-systems, such as neuraminidase or other glycosidases, lacto-peroxidase, and transglutaminase, which have access in intact cells exclusively to the exterior surface of the membrane, differential extraction with various solvents, and direct labeling with antibody − or lectin − ferritin conjugates. Most of these experimental approaches have been critically reviewed in several articles and the arguments will not be reproduced here [4, 6−8]. Despite some inconsistencies a concept emerged from these studies, which in a rather simplified way can be summarized as follows: (a) The bulk of the major proteins are located exclusively at the cytoplasmic side of the membrane, external to the lipid bilayer and can be classified as extrinsic proteins (polypeptide bands 1, 2, 4, 5, 6). (b) Two major proteins (glycophorin and polypeptide band 3) are accessible to probes at both sides of the plasma membrane and some portion(s) of the polypeptide chains has to be associated with the hydrophobic core of the bilayer because of this transmembrane

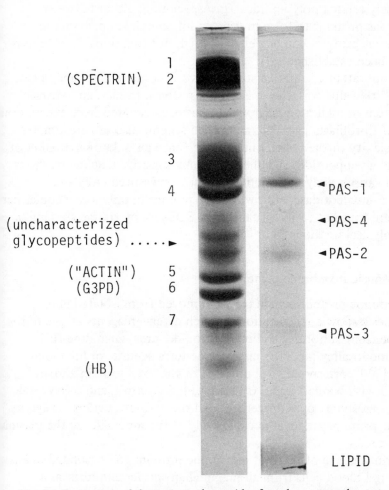

Fig. 3.1 Separation of the major polypeptides from human erythrocyte membranes by SDS-polyacrylamide gel electrophoresis. The numerical nomenclature is adopted from [5]. Gel on the left stained with Coomassie blue and on the right with periodic-acid-Schiff's (PAS) reagent. G3PD: glyceraldehyde-3-phosphate dehydrogenase, HB: hemoglobin.

configuration. (c) All the carbohydrate associated with the plasma membrane is located at the outer surface. (d) All the proteins known so far which are exposed to the outside are glycoproteins. (e) No major membrane protein is located exclusively at the outer surface of the membrane or is completely buried *within* the bilayer. (f) Each major band on SDS-gels probably represents a single major component or at least closely related polypeptides. (g) Associations either between homologous proteins to form multi-subunit complexes or between heterologous proteins can be demonstrated, but functional implications have not been established.

These are rather qualitative statements and some of them may prove to be of little value for other membrane systems. In fact, an extrinsic glycoprotein of high molecular weight has been isolated from the external surface of fibroblasts [9]. This simplified scheme also underestimates the complexity of the red cell membrane, since an unknown number of additional glycopeptides is demonstrable by sensitive labeling methods [10] and various enzymatic activities can be measured (ATPase, ACHase, 5'-nucleotidase, etc.) which are present in only a few copies per cell and are thus not easily visualized on SDS-gels by simple staining or other analytical methods [4].

3.2.1 Extrinsic membrane proteins

The membrane proteins which can be removed from isolated ghost membranes by low concentrations of such diverse reagents as *p*-chloromercuribenzene sulfonate, sodium hydroxide, urea, guanidine-HCl, lithium-diiodosalicylate, ethylenediamine-tetra acetate, or high ionic strength [11] correspond to about 40% of the total protein and are correlated with bands 1 and 2 (spectrin), 4, 5 ('actin'), and 6 (glyceraldehyde-3-phosphate dehydrogenase). All the available evidence suggests that these proteins are located exclusively at the inner side of the plasma membrane.

Spectrin was one of the first membrane proteins to be isolated and is found in red blood cells from all mammalian species and birds as a double band on SDS-gels. A crude preparation containing the two peptides of about 200 000–250 000 daltons (bands 1 and 2) and a peptide of 45 000 molecular weight (band 5) can easily be obtained and further purified. It has been difficult to separate the two high molecular weight peptides from each other despite the fact that the protein is quite soluble in low ionic strength buffers. However, the remarkable tendency of spectrin to aggregate in the presence of calcium and other metals or high salt concentrations has greatly complicated attempts to reveal further structural details. Cross-linking studies on ghost membranes indicate that the two peptides are close to each other or that they interact with each other in a fashion unknown at present. Although some evidence exists indicating the presence of several 'spectrins', no firm conclusions are possible as yet.

What is the role of these high molecular weight proteins? Guidotti suggested that the spectrin polymers resemble the large molecular weight polypeptides of muscle myosin and that they interact with actin or an actin-like protein (band 5) to form a contractile system for the membrane [12]. Such a system could function in some undefined way to maintain or change the shape of the red cells or to explain the endocytosis phenomena observed with isolated ghosts.

Singer and co-workers have provided some evidence for a relationship between spectrin and myosin [13]. By using antibodies prepared against smooth muscle myosin they found weak cross-reactivity with human spectrin. The model which they propose, visualizes spectrin polymers of varying sizes attached to the inner surface of the red cell at specific sites, the internal segments of transmembrane proteins. When the system is activated by unknown mechanisms, a direct polymerization reaction follows which results in the formation of rod-like or filamentous structures. The rearrangement of spectrin units may direct a redistribution of the transmembrane proteins to which they are anchored, thereby causing a corresponding change in the topography of their external receptor sites and establishing a new order at the surface of the cell. The reverse side of this model would be that interaction of ligands on the surface of the cell alters the distribution of the transmembrane receptor proteins which is followed by polymerization of spectrin subunits.

Spectrin has physicochemical characteristics which are related to those of myosin, but other properties, e.g. solubility or antigenicity are quite distinct. It is quite possible that the results described above are misleading and that there is only some myosin preserved in the red blood cell. Spectrin could be an entirely different protein, especially designed for the red cell, to make the membrane more resistant to the mechanical stress in the circulation. Similar high molecular weight proteins have been demonstrated in membrane preparations of phagocytic [14] and other cells, but there is no evidence that they are in fact related to spectrin, either chemically or functionally. More information is clearly needed, particularly in view of the fact that myosins from various non-muscle cells have distinct antigenic and chemical properties. The prevalent idea that spectrin forms an extensive filamentous network along the inner surface of the membrane is based on relatively indirect evidence. The filaments observed in some electron microscopic studies could be due to actin or the actin-like band 5 polypeptide and many attempts have failed to provide convincing evidence that spectrin itself forms such a network. Based on the observation that spectrin and actin interact *in vitro* and apparently inhibit polymer formation, other possibilities still have to be considered. Tilney and Detmers proposed that polypeptides 1, 2 and 5 are associated non-covalently and form a spectrin-actin complex. This complex does not polymerize to form extended filaments but rather builds up a 'mesh of unpolymerized material on the cytoplasmic surface of the lipid bilayer' [15].

3.2.2 Transmembrane proteins

Molecular models of biological membranes require an understanding of how proteins are inserted into the lipid bilayer, of how they interact with lipid molecules within the hydrophobic core, and of how protein–protein association in this environment is facilitated. The recently proposed fluid mosaic model is largely based on properties of the bilayer lipids and rather general characteristics of the proteins embedded into them. This model cannot make any predictions as to how individual protein molecules *are* interacting with the bilayer. Neither the molecular structure nor the lipid interactions of a single membrane protein are completely understood at present. Based on the few molecules which have been studied more extensively, one is rather forced to the conclusion that there is no single general mode of structure and conformation. However, at present one cannot be too dogmatic because of the rather limited information.

Two different types of transmembrane proteins have been identified. The first type are transport proteins, some of which as in the case of the ATPases, also possess enzymatic activity. A transport function may imply transmembrane configuration and this configuration is generally assumed. For plasma membrane ATPases, the 'receptors' for cardiac glycosides and ATP are located on different sides of the membrane. For the presumptive transport protein 'band 3' in red cell membranes, labeling and enzymatic digestion data suggest a transmembrane configuration.

The second type, which will be termed rather generally as 'receptor' proteins, has been extensively studied in the red cell membrane system. For one of these proteins, glycophorin A, the transmembrane configuration is firmly established.

Much of our interest in these proteins is focused on several questions:
(a) What proportion of the protein is buried within the lipid bilayer;
(b) What is the primary structure and conformation of the protein within the membrane which mediates interactions with lipids or other proteins;

(c) Is the functional part located outside or inside the membrane; (d) What is the significance of the cytoplasmic portion of these proteins; (e) How mobile are these proteins in the native unmodified membrane under physiological conditions; (f) How do they become inserted into the bilayer during biosynthesis.

The most prominent species of transmembrane proteins in the erythrocyte membrane are 'band 3' and glycophorin A (PAS-1) and these exist as dimers which extend completely across the bilayer.

Band 3 forms a rather diffuse band on SDS-gels and migrates as a polypeptide with a molecular weight of about 95 000. The peptide becomes labeled when radioactive probes impermeable* to the bilayer are added to intact cells. Portions of the polypeptide chain are removed when intact cells are incubated with pronase or chymotrypsin, reagents which cannot penetrate into the cell. The peptide can also be labeled by radioactive probes when introduced into ghost membranes or when they were applied to inverted plasma membrane vesicles. The important conclusions of many of these experiments are that band 3 has several domains, one of which is exposed on the exterior of the cell, one of which is demonstrable on the cytoplasmic surface, and, by inference, a third region which should be buried within the lipid bilayer.

Band 3 can be stabilized in a dimeric form when membranes are treated with bi-functional reagents or when oxidized by appropriate oxidizing conditions [4]. Since the band 3 polypeptide behaves homogeneously with regards to proteolytic cleavage [16], labeling, cross-linking, and isolation [17] the intriguing conclusion was that band 3 is a single homogeneous polypeptide. Intriguing, because several laboratories recently demonstrated good correlation between the inhibition of facilitated transport or diffusion of anions into the red cell and the covalent labeling of band 3 by certain reagents [18]. In addition, transport of *D*-glucose [19] and water [20] has been suggested to be mediated by band 3 polypeptides. Thus multiple functions are correlated with an apparently homogeneous peptide. This raises several possibilities, none of which can be disregarded: band 3 represents a family of closely related polypeptides and specialization occurs within the members of the family; band 3 is a homogeneous polypeptide, that is to say, the primary structure is homogeneous, but there are several functional sites on each protein; band 3 is a homogeneous polypeptide, but specialization for each transport site is conferred by different structures, e.g. degree of

* For an extensive discussion on this subject to recent reviews [4, 6–8].

glycosylation, or additional polypeptide subunits. As mentioned above, band 3 appears to be associated with spectrin and other major proteins like band 4 and band 6, the significance of which is not understood but which could provide discrimination.

Band 3 has been isolated and purified and found to contain 2–8% carbohydrates, mainly mannose, galactose, and *N*-acetyl glucosamine in approximate ratios 1:2:2, and traces of fucose and glucose [17, 21]. This composition suggests that the oligosaccharides are of the type which is linked *N*-glycosidically to asparaginyl residues. Mannose and galactose presumably are at the non-reducing ends of the oligosaccharide units, since concanavalin A and ricin bind to band 3. There is no evidence linking these 'lectin receptors' to any function of the protein, but their presence adds to the more general notion that all cell surface proteins may be glycosylated.

Preliminary work indicates that the size of the protein buried inside the lipid core of the membrane which is inaccessible to proteases from the inside and outside of the membrane amounts to only about 20% of the total mass [16]. This maximal value could still be a considerable overestimate. Thus it appears that at least some transmembrane proteins, even with more complex functions, may be in contact with lipid molecules with only a small portion of their polypeptide chains. From a linear arrangement of the polypeptide across the membrane one would predict that proteolytic treatment should release soluble peptides from the membrane. While this was found to be so for the cytoplasmic portion of band 3 when permeable ghost membranes were digested with trypsin or chymotrypsin, no fragment could be released from *intact* cells, that is, from the external surface with these enzymes. Trypsin apparently does not cleave the protein *in situ*, but chymotrypsin-treated cells yield two fragments of 35 000 and 55 000 on SDS-gels, yet attempts to release one of the fragments from the cell after cleavage were unsuccessful. This may suggest either that there is a very tight association of the exterior domain with the rest of the protein or that only the cleavage site is located exteriorly but the fragments remain associated with the membrane via polar or apolar interactions. Similar observations have been made with the ATPase from sarcoplasmic reticulum [22], and rhodopsin [23], and the possibility remains that a more complex structure, in which multiple sites along the polypeptide chains are interacting with the hydrophobic bilayer core, can account for these findings.

In a recent model of the purple membrane protein bacteriorhodopsin, a rather interesting structure emerged. The purple membrane of

Halobacterium halobium contains a single polypeptide chain of 20 000–26 000 molecular weight, which is thought to be completely buried in the bilayer [24]. X-ray diffraction patterns and image reconstruction from electron micrographs indicate that the protein is arranged in the membrane in a hexagonal array with three protein units clustered together [25, 26]. The model proposes that each polypeptide chain forms 7 α-helices in close association which traverse the bilayer at slightly different angles. According to this model, cleavage of accessible sites outside the bilayer would not release any part of the protein, since the peptides remain associated both with each other and with the lipid bilayer and only disruption of the membrane with detergents or other means would separate individual peptides.

3.2.3 The sialoglycoproteins (glycophorins)

On SDS-gels of human red cell membrane proteins four peptide bands are visualized with periodic acid–Schiff's reagent, a carbohydrate stain, which is particularly sensitive to the presence of sialic acid (Fig. 3.1). This pattern seems to depend upon the buffer system used during electrophoresis, since in some gel-systems only one band is observed with variable trace amounts of other bands. For some time it was difficult to decide whether the multiple bands represented true heterogeneity or were simply due to degradative artifacts created by proteolytic cleavage of one protein.

Original attempts to isolate the sialoglycopeptides were focused on the characterization of the receptors for influenza virus [27] and the molecule which carries MN-blood group activity [28].

Many investigators have tried to isolate these proteins in water-soluble form with varying degrees of success. Apparently similar protein fractions can be obtained by largely different extraction procedures such as phenol, pyridine, chloroform–methanol, *n*-butanol, non-ionic detergents, or lithium diiodosalicylate, but these preparations seem to vary significantly with regard to yield or size of the protein (reviewed in [8]. As we now know, as a true integral membrane protein the sialoglycoprotein or glycophorin as it has been termed later [29], probably derives its solubility in aqueous salt solutions mainly because of a considerable carbohydrate content, amounting to about 60% of its dry weight. In water, the protein apparently forms large aggregates of several hundred thousand molecular weight composed of an undetermined number of protein subunits (Fig. 3.2). Individual proteins seem to be aggregated via

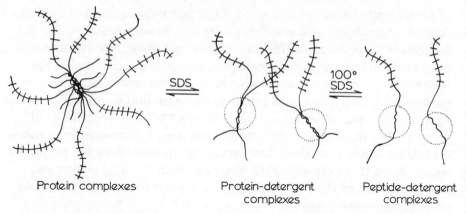

Protein complexes   Protein-detergent   Peptide-detergent
                               complexes   complexes

Fig. 3.2 Schematic representation of isolated membrane sialoglycoproteins
in aqueous salt solutions. The proteins form micellar structures in which
individual proteins associate via their respective hydrophobic regions. The
N-terminal glycopeptide and the hydrophilic *C*-terminal regions are located
at the exterior of the micelle and form the hydrophilic shell. Both regions
are accessible in the micelle to macromolecular probes such as proteases or
antibodies. The complex dissociates into individual dimeric proteins (PAS-1
form on SDS gels) in detergents. The monomeric subunit (PAS-2 form on
SDS gels) is obtained in the presence of high concentrations of SDS, low
ionic strength and low protein concentration at elevated temperature.
Both processes are reversible.

hydrophobic amino acid regions which are buried inside the micelle while
the water soluble regions, including the glycopeptide portions, form the
outer hydrophilic shell. Common denaturants like urea, guanidine-HCl,
high concentrations of chaotropic agents, or even organic solvents do not
or only partially reduce the size of these aggregates. These preparations
however, had high activity for inhibiting virus-, lectin-, or MN-antibody
mediated hemagglutination.

Molecular weight determination by gel filtration or ultracentrifugation
results in a wide range of values and even in the presence of detergents
the apparent monomer molecular weights range from 15 000 to 55 000.
Non-ionic and anionic detergents have been used successfully to reduce
the size of the protein micelles. Gel filtration in the presence of SDS or
non-ionic detergents like Triton X-100 or Ammonyx-LO clearly separates
the sialoglycoprotein fraction into several components, but none of the
individual peaks contains a single band as judged by SDS-electrophoresis
[30, 31]. The major fraction corresponds to 2 bands, PAS-1 and PAS-2,
on SDS-gels, and the additional protein peaks always yield bands at all

four positions of the starting material. Amino acid analyses revealed that these patterns were due to two different sialoglycoproteins, which were termed glycophorin A and glycophorin B, and that the separation achieved by gel filtration in detergents actually distinguishes glycoprotein-complexes composed of subunits in variable numbers. These two proteins are also isolated by affinity chromatography on wheat germ agglutinin-Sepharose in the presence of SDS and although the carbohydrate composition does not reveal differences, the difference in affinity is sufficient to separate the two proteins [32]. Accessibility of oligosaccharides to the covalently bound lectin or variation in the size of the glycoprotein complex with concomitant change in the affinity for the multivalent lectin may explain this property.

Glycophorin A has a molecular weight of 31 000. Its polypeptide chain is 131 amino acids long and, with the exception of two ambiguous positions, appears to be a homogeneous protein, at least with regard to the protein part ([33], Fig. 3.3). This contrasts with its behavior on SDS-gels, since two bands are obtained as mentioned before. This puzzle was solved when it was realized that the two peptide species represent interconvertible forms of the same protein ([31, 34], Fig. 3.2). This observation suggested either that PAS-1 was a multimeric form of PAS-2 or that PAS-2 was created by some conformational rearrangement or is due to a difference in SDS-binding. The conditions necessary for this conversion have been investigated in some detail and it appears now that the high molecular weight band, PAS-1, corresponds to the dimer of glycophorin A, but an even higher subunit number for this complex cannot be excluded. The subunits seem to be associated within the hydrophobic segment of the molecule (see below). The most interesting suggestion, however, was that this protein may form a similar complex in the plasma membrane. This would add an additional dimension to the complexity of this molecule, and could explain why MN-activity or lectin-binding activity are largely reduced when the glycopeptide segments of this protein are released from the cell, but cannot be completely recovered in the solubilized peptide fraction [35].

The amino acid sequence of glycophorin A is particularly interesting in that all of the oligosaccharide residues are attached to amino acids within the N-terminal third of the polypeptide chain (Fig. 3.3). Fifteen of the oligosaccharide units are linked to seryl (or threonyl) residues by O-glycosidic bonds and one more complex unit is linked N-glycosidic-ally to an asparaginyl residue. Different structures have been proposed for the carbohydrate chains: The structure I, proposed by Springer and

Leu-Ser*-Thr-Thr*-Glu-Val-Ala-Met-His-Thr*-Thr*-Thr*-Ser-Ser*-Ser-Val-Ser-Lys-Ser-Tyr-
Ser   Gly   10   20

Ile-Ser*-Ser-Gln-Thr*-Asn**-Asp-Thr-His-Lys-Arg-Asp-Thr-Tyr-Ala-Ala-Thr*-Pro-Arg-Ala-
                                          30   40

His-Glu-Val-Ser-Glu-Ile-Ser-Val-Arg-Thr*-Val-Tyr-Pro-Pro-Glu-Glu-Glu-Thr-Gly-Glu-
                         50   60

Arg-Val-Gln-Leu-Ala-His-His-Phe-Ser-Glu-Pro-Glu-Ile-Thr-Leu-Ile-Ile-Phe-Gly-Val-
                             70   <u>80</u>

<u>Met-Ala-Gly-Val-Ile-Gly-Thr-Ile-Leu-Leu-Ile-Ser-Tyr-Gly-Ile-Arg-Arg-Leu-Ile-Lys-</u>
                             90   100

Lys-Ser-Pro-Ser-Asp-Val-Lys-Pro-Leu-Pro-Ser-Pro-Asp-Thr-Asp-Val-Pro-Leu-Ser-Ser-
                             110   120

Val-Glu-Ile-Glu-Asn-Pro-Glu-Thr-Ser-Asp-Gln-COOH
                             130

Fig. 3.3 Amino acid sequence of glycophorin A from human erythrocyte membranes. Heterogeneity in positions 1 and 5 was found and has not been explained yet. * denotes O-glycosidically linked oligosaccharide, ** denotes one N-glycosidically linked oligosaccharide. The underlined sequence from position 73 to 95 indicates the presumptive length of the hydrophobic segment. The exact number of residues which are buried within the bilayer is however not known. The sequence from position 102 to 118 is the maximum size of an antigenic determinant, which is found at the cytoplasmic side of the membrane.

I   NANA $\xrightarrow{\alpha 2-3}$ Gal $\xrightarrow{\beta 1-3}$ GalNac $\longrightarrow$ Ser(or Thr)

$$\uparrow \alpha 2-6$$

NANA

II   NANA $\xrightarrow{\alpha}$ Gal $\xrightarrow{\beta}$ GalNac $\xrightarrow{\alpha_1}$ Ser(or Thr)

$$\uparrow \beta$$

NANA $\xrightarrow{\alpha}$ Gal

III   NANA $\xrightarrow{\alpha}$ Gal $\xrightarrow{\beta}$ GlcNAc $\xrightarrow{\beta}$ Man $\xrightarrow{\alpha}$ Man $\xrightarrow{\alpha}$ GlcNAc $\xrightarrow{\beta}$ Asn

$$\uparrow \beta$$

NANA $\xrightarrow{\alpha}$ Gal $\xrightarrow{\beta}$ GlcNAc

IV   NANA $\xrightarrow{\alpha}$ Gal $\xrightarrow{\beta}$ GlcNAc

$$\searrow \beta$$

Gal $\xrightarrow{\beta}$ GlcNAc $\xrightarrow{\beta}$ (Man)$_3$ $\xrightarrow{?}$ GlcNAc $\xrightarrow{\beta}$ Asn

$$\nearrow \beta$$

Fuc $\xrightarrow{\alpha}$ Gal $\xrightarrow{\beta}$ GlcNAc

Desai [36] is reported to have MN-antigenic activity and differs from II [37] by an additional galactose residue. Carbohydrate compositions of glycopeptides prepared from glycophorin A favor structure II since equivalent amounts of galactose and GalNAc were found. However it is possible that some of the oligosaccharides have the alternate structure. From work on fetuin it is known that the NANA linked to GalNAc is possibly found only in oligosaccharides attached to seryl residues but not when linked to threonine [38].

Further work will be also required to distinguish between the conflicting structures III [39] and IV [40]. Since these carbohydrate sequences have been obtained from peptide mixtures misleading results could have been obtained. In soluble proteins in some instances asparaginyl residues found in the general sequence Asn-X-Ser (Thr) are always glycosylated.

In glycophorin A, the only asparagine found in this sequence is glycosylated. It is interesting to note that the two different types of oligosaccharides which are found on glycophorin A are similar to those found on many other glycoproteins. Thus, if carbohydrate moieties on glycophorin contribute significantly to any unique antigenic structure or any special receptor, they probably do so on the basis of distribution or variation which may be under control of the primary structure of the protein or on the basis of some co-ordinated interaction with the polypeptide backbone. It has been postulated that amino acid side chains contribute or are necessary for the antigenic expression of MN-activity [41].

The *C*-terminal end of glycophorin is particularly rich in proline residues and acidic amino acids. This part of the molecule is thus well suited to bind cations, such as calcium, or basic peptides. Thus both the *N*-terminal glycosylated half of the molecule and the *C*-terminal acidic region would be capable of interacting with hydrophilic moieties at both surfaces of the membrane.

The segment of the molecule which interacts with the lipid bilayer was originally identified on the basis of its insolubility in aqueous solutions. It consists of 23 residues many of which are hydrophobic [42]. Evidence that glycophorin A subunits associate within this region is mainly derived from studies on the behavior in SDS by SDS gel electrophoresis [31]. It can be demonstrated that the tryptic peptide containing the hydrophobic region of glycophorin binds to the presumptive subunit (PAS-2) even in SDS but not to the oligomeric form (PAS-1). Modification of a single methionyl residue within the hydrophobic region leads to a change in the peptide, and binding is no longer observed [43]. It should be pointed out, however, that from sedimentation equilibrium studies in the presence of SDS, a molecular weights of 29 000 has been calculated for glycophorin A [44]. There is no ready explanation for this result since, under conditions in which the analysis was done, most of the glycoprotein was predicted to be in the dimeric form. The sialoglycoprotein from human red cell membranes as well as other membrane proteins containing hydrophobic regions has been shown to bind higher amounts of SDS than soluble proteins and also binds non-ionic detergents and deoxycholate, which are usually not bound to soluble proteins. In view of these anomalies and the limited experience available with membrane proteins, interpretations of such data have to be made rather cautiously and may still be somewhat provisional at this time.

Glycophorin B has been studied less extensively. Upon re-electrophoresis the material isolated from the PAS-3 position on SDS gels reveals multiple bands which co-migrate with the PAS-1, -2, -3, and -4 bands and which fit an oligomeric series. Similar aggregation was observed for glycophorin B, isolated as described above [30]. Preliminary studies on proteolytically derived peptides of glycophorin B indicate that the overall structure is very similar to glycophorin A. The amino terminal region contains all the carbohydrate found on the molecule, and some regions are identical in amino acid sequence to glycophorin A. A hydrophobic peptide, similar to the region found in glycophorin A should serve to anchor this protein in the bilayer. It was interesting however, that the C-terminal peptide, 35 amino acids long in glycophorin A, was not demonstrable for glycophorin B. In support of this conclusion is an immunochemical study [45]. Isolated antibodies to a *C*-terminal region (residues 102–118) did not react with glycophorin B. This could be interpreted as a lack of cross-reactivity between the two sialoglycoproteins due to differences in primary structure, but the absence of a *C*-terminal region or a peptide corresponding to it strongly favors a shortening of glycophorin B as compared to A at the *C*-terminal end.

This raises the interesting question about the function and importance of the cytoplasmic extension of transmembrane proteins in intact cells. Although proteolytic removal during preparation of the peptide has to be considered, this is not a likely explanation. The membrane of the intact red cell does not readily fit the 'fluid mosaic' model since major integral membrane proteins like band 3 or glycophorin seem to have a fixed location and have not been shown to exhibit lateral diffusion. Removal of spectrin by extraction or tryptic digestion of isolated ghost membranes causes redistribution of intra-membranous particles in freeze-cleaved membranes, which are thought to be correlated with the two major transmembrane proteins. Such lateral mobility would be expected to depend primarily on the viscosity of the lipid hydrocarbon core and on the net electric charge occurring on the portion of the molecule protruding into the aqueous surface of the membrane. It may also depend on nearest neighbor interactions and the shielding of this electric charge by counter-ions. The hydrophilic portions of transmembrane proteins thus may provide structures which determine the planar topographical arrangement by interaction with other proteins. Removal of this restraint allows these proteins to undergo translational motion. The absence of an extended cytoplasmic polypeptide region in glycophorin B could indicate that it is not connected or entrapped within the spectrin meshwork and thus may be expected to be mobile. However, it is possible that this protein is complexed to other polypeptides in the membrane or that other properties of the bilayer restrict the motion. Some membrane proteins, which do not appear to have extended cytoplasmic regions, have been shown to exhibit lateral mobility in studies of the diffusion of fluorescent antibody or lectin-labeled membrane antigens over the surface of the membrane. These qualitative observations deal mainly with the diffusion of membrane protein-ligand complexes and thus severe perturbation of the membrane environment under these *in vitro* conditions cannot be excluded.

The transmembrane orientation of the major sialoglycoprotein has been demonstrated repeatedly in many laboratories throughout the world primarily on the basis of differential labeling of the molecules in intact cells versus leaky ghost membranes (for reviews see [4, 6–8]). Since objections have been raised to this approach, several investigations have attempted to show that the presumptive cytoplasmic segment of glycophorin can be labeled by introducing labeling reagents inside ghost membranes or labeling inside-out vesicles. Both approaches seem to have been successful. However other studies have failed to label sialoglycoproteins by this method. Lactoperoxidase-catalyzed iodination mainly labels tyrosine residues, and from the amino acid sequence of glycophorin it is known that only a single tyrosine residue is located at the *C*-terminal end of the hydrophobic region. This tyrosine may still reside within the hydrocarbon core of the bilayer. The variability in reactivity of this tyrosine may be understandable, if it depends on experimental conditions such as preparation of membranes or excess generation of ^{125}I radicals, which produces long-range labeling despite inaccessibility of the site to lactoperoxidase.

The most convincing evidence for a transmembrane orientation of glycophorin A comes from electron microscopic studies on intact cells [45]. Highly specific antibodies to an antigenic site on the *C*-terminal region of glycophorin when coupled to ferritin, localize exclusively inside the cytoplasm close to the membrane on frozen thin sections. The electron-dense particles are regularly distributed throughout the circumference of the membrane which is expected from the topographical distribution of the polypeptides in the membrane. Interestingly, the particles are found at equal distances from the membrane, which seem to reflect the linear mode of insertion into the membrane of the peptide portion of glycophorin. Since this interaction is dependent upon and fixed by the hydrophobic region, one might expect that the *C*-terminal region has a similar fixed location. In contrast, conjugates prepared from the plant lectin wheat germ agglutinin localize exterior to the membrane, but at various distances. Wheat germ agglutinin receptor activity is associated with the sialoglycoproteins. The uneven distribution of this marker on the exterior surface could indicate, that glycophorins assume different conformations. It is conceivable that heterogeneity of the carbohydrates has a profound effect on the tertiary structure and as a result the markers are seen at different distances from the membrane. Alternatively, the dense packing of the oligosaccharides along the poly-peptide backbone allows only a rod-like extended structure, which protrudes from the surface of the cell. In this linear model of glycophorin, receptors may be located on separate molecules, e.g. glycophorin A and B, but occupy different positions on the polypeptide chains.

3.3 LIPID-PROTEIN INTERACTION

Hydrophobic substances are defined as molecules that are readily soluble in polar solvents but only sparingly in water [46]. Attraction of non-polar groups for each other plays only a minor role in this phenomenon. The 'hydrophobic effect' is rather due to the strong attractive forces between water molecules which must be disrupted when any solute is dissolved in the water. If the solute is ionic or strongly polar, it can form strong bonds to water molecules, which compensate for the disruption or distortion of the bonds existing in pure water; and ionic or polar sub-stances thus tend to be easily soluble in water. No such compensation occurs with non-polar groups and their solution in water is accordingly resisted. In fact they tend to cluster together by excluding water and thus decrease the free energy of the system. Molecules which are amphiphilic, with an ionic part soluble in water and a second part expelled from water as a result of hydrophobicity, will be forced to adopt unique orientations and to adopt certain organized structures. This is the case in biological membranes, which can form spontaneously and in which individual amphiphilic lipids are arranged in a bilayer structure. The polar head groups interact with water, while the hydrocarbon tails are segregated into the interior of the membrane.

From the previous discussion on erythrocyte membrane proteins, it is clear that some proteins have to interact with the hydrocarbon region of the bilayer, since they span the entire thickness of the membrane with polar regions extending into the aqueous environment at both sides of the membrane. For this reason these proteins have been compared with the more simple amphiphilic molecules. How do these proteins interact with lipid molecules and what is the structural basis for this interaction? Is there a unique protein structure, which is common to all these proteins or are there multiple forms of organization, which are suited for the hydrophobic environment of the bilayer membrane?

Very few membrane proteins have been studied in enough detail, but some concepts seem to have emerged recently. The most extensively studied protein in bacterial membranes is a small molecular weight lipoprotein which apparently is almost completely buried in the outer membrane [47]. It can be isolated from the cell envelope of *E. coli* by boiling SDS together with the peptidoglycan layer, to which it is linked covalently. In addition, the outer membrane contains about twice as much lipoprotein in a free form, which is not linked to the peptidoglycan. The protein is 58 amino acid residues long and surprisingly is rather polar but contains several extraordinary features: (a) starting with asparagine in the fourth position, non-polar residues occur at regular intervals of every third and fourth position. In other words, the distribution exhibits a regular pattern in two series, in both of which hydrophobic residues occupy every seventh position. This distribution has been found also in the *C*-terminal half of tropomyosin as well as in other proteins; (b) the *N*-terminal cysteine is modified to glyceryl cysteine and contains two fatty acids in ester linkage and one fatty acid in amide linkage; (c) the ε-amino group of the *C*-terminal lysyl residue is linked to the murein subunit; (d) there are amino acid sequence duplications.

The lipoprotein has an α-helical content of about 80% which is reduced by about 50% in SDS solutions. Based on the distribution of the hydrophobic residues and the similarity to the tropomyosin molecule a regular α-helical model was suggested. As in tropomyosin, this arrangement leads to the alignment of all hydrophobic side chains on one face of the helical rod. In a more detailed model, Inouye proposed that six or more of these α-helical rods are assembled to form a 'superhelix' stabilized by ionic interactions [48]. The hydrophobic residues in this model are facing outside and interact with hydrocarbon chains of lipids in the bilayer. In addition the covalently linked fatty acids are flipped back and are inserted into the outer half of the bilayer to provide additional stability. This assembly could span the entire thickness of the membrane and would have the property of a hydrophilic, acidic channel with a pore size of 12.5 Å, permitting the passage of many small molecules through the outer layer.

The second protein or class of proteins, the amino acid sequence of which is entirely known, is the coat protein of the filamentous bacteriophages of *E. coli* [49]. Studies indicate that this protein becomes inserted into the bacterial membrane after infection or during *de novo* synthesis and thus can be considered an integral membrane protein in the host, although the coat of the mature virus does not contain lipid and is made up almost completely of this protein of 5000 molecular weight. This protein which has been isolated from various phages does, in fact, contain a cluster of about 20 predominantly hydrophobic amino acid residues. This hydrophobic region is situated between an acidic *N*-terminal and a basic *C*-terminal segment [50]. It is assumed that the protein has a transmembrane configuration in the bacterial membrane, with the *N*-terminus located at the exterior surface. Data on the incorporation of the protein into artificial bilayer vesicles are in support of this orientation [51]. This model resembles very much the model proposed for glycophorin A in human erythrocyte membranes. As indicated in low resolution X-ray diffraction studies, in the lipid-free virus coat the protein is arranged as helical rods that extend axially and overlap each other like shingles to form a cylindrical shell containing the DNA [52, 53]. Circular dichroism measurements support a high helical content of more than 90%. The conformation of the protein in the bacterial membrane is unknown, but model studies suggest that the protein in detergents or phospholipid micelles forms dimers and undergoes a considerable conformational change. There is a loss of α-helix as indicated by CD-measurements and a structure more resembling β-pleated sheet was observed [54]. This result is in contrast to similar studies on the hydrophobic peptide of glycophorin A, which showed a high helical content even in the presence of detergents [42].

A variety of viruses in animals, plants, or bacteria contain a limiting bilayer membrane termed the viral envelope. Extensive studies have demonstrated that only a few polypeptides are associated with the viral membranes, which are inserted into the bilayer on the exterior or interior side. Some of the glycoproteins which are found protruding from the surface of a variety of viruses as spikes, can be removed by proteolytic digestion without disrupting the integrity of the virus particles or the membrane (for review see [55, 56]. Trypsin releases neuraminidase and erythrocyte-binding activities into the supernate after limited digestion of influenza or parainfluenza virus, but the released fragments do not agglutinate erythrocytes. This suggests that the agglutinin on the virus acts as a multivalent ligand. Isolated peptides derived from the virus hemagglutinin bind, but cannot agglutinate cells any longer. Disruption of the virus membrane by SDS-solutions allows the isolation of intact molecules which also bind to erythrocytes without causing hemagglutination. Removal of the detergent, however, causes an increase in the sedimentation coefficient, indicating that the protein aggregates in the absence of detergent, presumably via a hydrophobic peptide region. In this aggregated form the protein is a potent hemagglutinating agent. These observations are consistent with the idea, that these glycoproteins, expressed on the surface of virus envelopes, do have peptide regions which serve as anchors to insert them in the membrane. After tryptic digestion of intact virus particles these peptide regions remain associated with the membrane. Disruption of the bilayer is needed to recover the whole intact protein, but their physicochemical and biological activities are very much dependent on the state of aggregation.

More direct evidence was provided by studies on Semliki Forest virus, a group A arbovirus, which contains two glycoproteins with an apparent molecular weight of 50 000—55 000. Digestion of the intact virus with thermolysin removes the spikes which results in a loss of hemagglutinating activity and infectivity. SDS-electrophoresis shows that after proteolytic treatment the membrane glycoproteins disappear, and instead, small molecular weight peptides are found which are considerably hydrophobic as indicated by their solubility in detergents, organic solvents, and by their amino acid composition [57].

An apparently similar arrangement has been found for the enzyme cytochrome b_5 from endoplasmatic reticulum [58]. Proteolytic digestion removes enzymatic activity from the membrane particles and a slightly smaller protein is recovered, which has full activity. The small fragment which remains associated with the membrane contains a high number of hydrophobic residues. Although amino acid sequence data are not available the amino acid composition suggests a linear hydrophobic sequence.

These rather limited studies may indicate that membrane-associated polypeptide regions have amino acid sequences rich in hydrophobic amino acids which are not interrupted by charged residues. This particular arrangement may be found in membrane proteins the major function of which resides on portion of the molecule exterior to the membrane. Such functions may include: to serve as receptor and recognition units or as enzymes or structural molecules capable of interacting with extrinsic proteins. The rather short, 20—25 amino acid long sequences serve to firmly bind the protein to the membrane and, besides being suited for interaction with the hydrocarbon tails of the lipid molecules, these regions may form stable or reversible associations with similar regions of other polypeptide chains. The *E. coli* lipoprotein studies alternatively show that linear hydrophobic sequences may not be a structural prerequisite for lipid interaction.

In earlier studies it has been concluded that membrane proteins are likely to have an α-helical conformation [59]. O.R.D. and C.D. spectra of red cell ghosts suggested that a substantial portion of the proteins was in the α-helical conformation. Certain anomalies in these spectra were also interpreted in favor of the idea that most of the polypeptides in the α-helical conformation were in close proximity to the hydrocarbon segments of phospholipids. But these anomalous spectra are probably due to light scattering effects rather than to some special environment of the helical peptide regions. Studies discussed earlier using enzymatic probes, labeling techniques or extraction methods on red cell membranes find only a small number of polypeptides in association within the lipid bilayer, and it is questionable whether the high helix content extimated from the spectra can be used to make predictions about protein conformation within the hydrophobic core in complex membrane systems. Individual proteins or protein complexes have been successfully reconstituted with lipid and restoration of function has been obtained. Analysis of isolated membrane proteins or fragments in some of the detergents or after incorporation into phospholipid membranes by these physical methods appears to be more promising now, despite the possible modification of protein conformation by the methods commonly used to isolate and purify membrane components. Although a high α-helical content has been suggested for a small number of isolated membrane proteins or hydrophobic peptides of such proteins in more recent studies, it still seems premature to conclude that this is the only conformation proteins can assume within the bilayer or which is required for interaction with the hydrocarbon chains of the phospholipids.

Amino acid compositional data for many 'integral' membrane proteins, operationally defined as such by criteria such as solubility or reconstitution experiments, are usually not very reliable or sufficient indicators to distinguish between water-soluble protein and membrane proteins. Attempts have been made to calculate the polarity or hydrophobicity of proteins from the amino acid composition, and from the polarity or hydrophobicity of various amino acids, to establish criteria for membrane proteins [60]. These attempts were only occasionally successful. This may be due to the fact that large membrane proteins have only relatively short hydrophobic sequences, or alternatively, that a specialized but different structural arrangement may allow these proteins to interact with lipids, both of which would not be discovered by these calculations. The second alternative may be more frequently found associated with complex systems such as transport proteins.

It remains to be seen whether these rather simple models can provide the structural basis for the diversity of functions which are associated with membrane proteins. One can only speculate, that various structures and forms of organization will be found, as structural details of more membrane proteins will become available. The principle of short peptide regions which span the lipid bilayer may be used, for instance, to accomodate larger portions of proteins within the membrane. Multiple short segments of the same polypeptide chain, connected by loops, may traverse the bilayer as suggested by the studies on the purple membrane protein. Alternatively, there may be a dissociation in structure. There may be small proteins or peptides which reside solely within the lipid bilayer to form multi-subunit complexes which would be well-suited to serve as channels or transport sites. This complex would not be able to regulate transport of molecules across the membrane itself, but would derive this function from different subunits associated with it exterior to the membrane. This form of organization may be the underlying principle for multi-subunit membrane components such as the mitochondrial ATPase [61] or other transport proteins.

There is no evidence yet for large globular protein structures within membranes, although the so called intramembranous particles observed on freeze-fracture images of plasma membranes have been interpreted as such. There is agreement that freeze-cleavage exposes the hydrophobic core of the lipid bilayer by removing one-half of the membrane. The globular structures are due at least in part to the presence of proteins in the bilayer, since no particles are seen on freeze-fractured artificial bilayer membranes. The approximate dimensions of the particles are in the 70–80 Å range, but these measurements are done on platinum-carbon replicas, which make it difficult to determine the mass of the protein. The current interpretations are, that these particles correspond to intramembranous regions, protein complexes, or lipid-protein complexes of integral membrane proteins. But it is not clear what segments of such proteins are visualized. Conceivably this technique does not distinguish between different regions interior or exterior of the membrane and the particles may contain a larger mass than is accounted for by intra-membranous polypeptide segments.

3.4 GLYCOPROTEINS IN RED CELL MEMBRANES FROM OTHER SPECIES

The molecules at the surface of cell membranes in mammalian species seemingly serve a critical function in carrying markers of individuality, blood group antigens, or histocompatibility antigens, as well as markers which distinguish various cells from each other. In lymphocytes, histocompatibility antigens are coded for, and are closely linked to, genes which determine the immune response to foreign material, and seem to play an important role in cell interactions. The close linkage between genes determining cell surface molecules to the complex intracellular functions of highly specialized cells such as the lymphocytes, has suggested that the specific immune system has evolved from more primitive recognition and defense mechanisms in which the plasma membrane played the decisive role. One would predict from these considerations, that membrane proteins involved in cell-cell recognition, contact maintenance or as markers of individuality have undergone greater changes during evolution than membrane proteins important for metabolism, transport, or motility.

Comparative studies on red cell membranes of different mammalian species, in fact, show that most major proteins or polypeptides as described in the previous sections are remarkably similar by their mobility on SDS gels, antigenic properties, and amino acid composition. In contrast, the major sialoglycoproteins from red cell membranes of several mammalian species and birds [62] have quite distinct SDS gel profiles. Some species contain a single major glycopeptide band or have in addition a few minor bands, while others do not show major bands at all but contain instead a series of minor components or very little PAS-stainable material. The PAS-reagent detects predominantly sialic acid and cannot really give a glycoprotein profile, particularly since the sialic acid content of different mammalian species is highly variable. In addition, the interpretation of such results can be misleading due to the aggregation phenomena described earlier. Isolation and preliminary characterization by amino acid and carbohydrate composition of glycoprotein (mixtures) of various species however, tend to support the initial impression that there is a considerable variation. A considerable difference was found even for two different breeds of swine [63]. Thus, large differences in amino acid and carbohydrate composition appear to be correlated with the differences in the SDS-gel patterns. The basis for the heterogeneity seen on SDS gels for most species however remains to be determined. The differences in mobility may be due to the large variation in carbohydrate content in addition to protein complex formation. It is possible that at least the basic principle was preserved for glycoproteins from red cell membranes, namely that of a rather small protein, which is heavily glycosylated. The amino acid sequence may determine the location, distribution, and the quantitative and qualitative degree of glycosylation, which determine to a large extent serological and other activities.

3.5 SEROLOGICAL ACTIVITIES ASSOCIATED WITH SIALOGLYCOPROTEINS

Several antigenic activities are associated with the red cell glycoproteins. In humans, MN blood group activity was only found in the cell-bound state and not in soluble form in serum or in secretions [64]. It appears that MN is found only on the major glycoprotein (glycophorin A) but not on the minor glycoprotein species. This is somewhat surprising in view of the ubiquitous nature of the serine (threonine) type oligosaccharide on these proteins. I, S, and ABH activity on the other hand are associated with the minor sialoglycoprotein, presumably corresponding to glycophorin B [65]. It is clear that ABH substances are present on both, glycoproteins from various secretions and cell membrane glycolipids and several variants have been isolated and their structure determined [64]. The structure and location of these carbohydrate determinants on cell membrane glycoproteins is not known. ABH activity is found on many other cell types, including endothelial and epithelial cells, and other blood cells. It is possible that some of the glycoproteins found on red cells have a wider distribution amongst various tissues. Sialoglycoproteins have been demonstrated in a great many other cells, but no information is available to confirm or discard the possibility that glycophorin A or an altered form of it is present, for example, on fat cells. It seems clear, however, that the complex structure of these molecules is not exhausted by the few serological activities analyzed so far. ABH or MN-activity is probably associated with only a few oligosaccharide structures on each protein. What is the function of these complex molecules in red cell membranes as well as on other cell surfaces? This question still eludes us and it will be reserved to future efforts to unravel the biological implications.

3.6 SUMMARY

The red blood cell is an end cell and has provided us with a unique opportunity to study structural details of a reasonably complex mammalian membrane system. As an end-cell without active protein synthesis it cannot respond to environmental stimuli like 'living' cells with *de novo* synthesis of various secretory products or modulation of the cell surface and this is one of the major assets of this system and has greatly facilitated studies on the arrangement of the major membrane proteins. Phospholipid bilayers do not determine the main features of the gross mechanical properties of red cells and probably most other cells. A macromolecular network of large molecular weight proteins, located at the cytoplasmic side of the membrane may be primarily responsible for maintaining shape and mechanical viscosity properties. Lateral diffusion within the plane of the membrane of proteins inserted into the bilayer has not been observed in erythrocyte membranes and we speculate that the macromolecular network restricts lateral mobility possibly by direct non-covalent interaction. In other cells such submembranous structures may serve additional functions as anchors for contractile elements important for motility or to direct the distribution of receptors at the cell surface. How this is achieved by the cell is still largely open to speculation.

Some of the membrane proteins are integrated into the lipid bilayer and we are just beginning to understand the structural requirements and possible modes of this interaction. Various models have been presented for some of the proteins under study but no definitive answers are available yet. There is however a strong suggestion, that helical rod-like structures, spanning both halves of the bilayer, are common to at least some of the membrane proteins. These may be found in different arrangements by forming multi-subunit aggregates to produce channels or pores through the membrane or complex 'receptor' structures on the cell surface. Many questions still have to be answered. What forces keep individual proteins apart from each other? Is this a property intrinsic to the lipid bilayer or are there external mechanisms which control their positions? Can hydrophobic proteins undergo reversible interactions within the apolar environment of the membrane? It seems clear that these problems become approachable as more information becomes available in this rapidly moving field.

One of the more interesting problems is posed by the complex glycoprotein structures seemingly present on many if not all cell surfaces.

What is the functional role of these macromolecules? The specific inter-action with lectins, antibodies, viruses, or other parasites may be an expression of a profound relationship between these marcomolecules and other soluble mediators, organisms, or cells. Carbohydrates certainly have a profound influence on the shape of the molecule depending on the number, distribution, and type of sugar and linkages. Oligosaccharide structures may have a considerable influence on the maintenance of tertiary structure of the protein and in addition may protect against proteolytic attack, similar to the subtle alterations in conformation engendered in soluble proteins by suitable modification of amino acid side chains. It is conceivable that such molecules could serve as cell recognition sites complementary to sites on other cells. The negative charge on the surface of the red cell as well as on other cells is largely due to sialic acid. The considerable variability in sialic acid content between various species may suggest, that parallel changes occur on cells other than the red cell, e.g. endothelial cells, in order to 'balance' the system. It would be interesting to comparatively analyze other cell types within one species. Modification of terminal carbohydrates, e.g. removal of sialic acid residues, will be recognized by the environment, as studies on serum glycoproteins suggest. In mammals a 'lectin'-like protein in the liver, specific for a certain arrangement of exposed galactose residues, will remove asialoglycoproteins from the serum. In birds, however, this liver receptor has specificity for terminal *N*-acetyl-glucosamine residues [67]. The high levels of circulating asialoglyco-proteins in the serum of birds apparently made a change in the specificity of the receptor protein in the liver mandatory.

REFERENCES

1. Bennett, D., Boyse, E.A. and Old, L.J. (1972), In: 'Cell Interaction', *Proc. Third Lepetit Colloquium 1971* (Silvestri, L.G., Ed.), pp. 247–263. North Holland Publishing Company, Amsterdam.
2. Marks, P.A., Rifkind, R.A. and Bank, A. (1975), In: 'Biochemistry of Cell Differentiation' (Paul, J., Ed.), *MTP Int. Rev. Sci.* **9**, 129–160.
3. Singer, S.J. and Nicolson, G.L. (1972), *Science,* **175**, 720–731.
4. Steck, T.L. (1974), *J. Cell Biol.,* **62**, 1–19.
5. Fairbanks, G., Steck, T.L. and Wallach, D.F.H. (1971), *Biochemistry,* **10**, 2606–2617.
6. Bretscher, M.S. (1973), *Science,* **181**, 622–629.
7. Carraway, K.L. (1975), *Biochim. biophys. Acta,* **415**, 379–410.

8. Marchesi, V.T., Furthmayr, H. and Tomita, M. (1976), *Ann. Rev. Biochem.* **45**, 667–698.

9. Yamada, K.M. and Weston, J.A. (1974), *Proc. natn. Acad. Sci. U.S.A.*, **71**, 3492–3496.

10. Gahmberg, C.G. (1976), *J. biol. Chem.* **251**, 510–515.

11. Steck, T.L., Yu. J. (1973), *J. supramolec. Struct.* **1**, 220–232.

12. Guidotti, G. (1972), *Ann. Rev. Biochem.*, **41**, 731–752.

13. Painter, R.G., Sheetz, M. and Singer, S.J. (1975), *Proc. natn. Acad. Sci. U.S.A.*, **72**, 1359–1363.

14. Hartwig, J.H. and Stossel, T.P. (1975), *J. biol. Chem.*, **250**, 5696–5705.

15. Tilney, L.G., and Detmers, P. (1975), *J. Cell Biol.*, **66**, 508–520.

16. Steck, T.L., Ramos, B. and Strapazon, E. (1976), *Biochemistry*, **15**, 1154–1161.

17. Furthmayr, H., Kahane, I. and Marchesi, V.T. (1976), *J. Mem. Biol.*, **26**, 173–187.

18. Rothstein, A., Cabantchik, Z.I. and Knauf, P. (1976), *Fed. Proc.*, **35**, 3–10.

19. Lin, S. and Spudich, J. (1974), *J. biol. Chem.*, **249**, 5778–5783.

20. Brown, P.A., Feinstein, M. B. and Sha'afi, R.I. (1975), *Nature*, **254**, 523–525.

21. Tanner, M.J.A. and Boxer, D.H. (1972), *Biochem. J.*, **129**, 333–347.

22. MacLennan, D.H. (1975), *Can. J. Biochem.*, **53**, 251–261.

23. Pober, J. and Stryer, L. (1975), *J. mol. Biol.*, **95**, 477–481.

24. Oesterhelt, D. and Stoeckenius, W. (1971), *Nature New Biol.*, **233**, 149–151.

25. Unwin, P.N.T. and Henderson, R. (1975), *J. mol. Biol.*, **94**, 425–440.

26. Henderson, R. and Unwin, P.N.T. (1975), *Nature*, **257**, 28–32.

27. Kathan, R.H., Winzler, R.J. and Johnson, C.A. (1961), *J. exp. Med.*, **113**, 37–45.

28. Springer, G.F., Nagai, Y. and Tegtmeyer, H. (1966), *Biochemistry*, **5**, 3254–3272.

29. Marchesi, V.T., Tillack, T.W., Jackson, R.L., Segrest, J.P. and Scott, R.E. (1972), *Proc. natn. Acad. Sci. U.S.A.*, **69**, 1445–1449.

30. Furthmayr, H., Tomita, M. and Marchesi, V.T. (1975), *Biochem. Biophys. Res. Commun.*, **65**, 113–121.

31. Furthmayr, H. and Marchesi, V.T. (1976), *Biochemistry*, **15**, 1137–1144.

32. Kahane, I., Furthmayr, H. and Marchesi, V.T. (1976), *Biochim. Biophys. Acta*, **426**, 464–476.

33. Tomita, M. and Marchesi, V.T. (1975), *Proc. natn. Acad. Sci. U.S.A.*, **72**, 2964–2968.

34. Marton, L.S.G. and Garvin, J.E. (1973), *Biochem. Biophys. Res. Commun.*, **52**, 1457–1462.

35. Adair, W.L. and Kornfeld, S. (1974), *J. biol. Chem.*, **249**, 4696–4704.

36. Springer, G.F. and Desai, P.R. (1974), *Biochem. Biophys. Res. Commun.*, **61**, 470–475.

37. Thomas, D.B. and Winzler, R.J. (1969), *J. biol. Chem.*, **244**, 5943–5946.

38. Spiro, R.G. and Bhoyrog, V.D. (1974), *J. Biol. Chem.*, **249**, 5704–5717.

39. Kornfeld, R. and Kornfeld, S. (1970), *J. biol. Chem.*, **245**, 2536–2545.

40. Thomas, D.B. and Winzler, R.J. (1971), *Biochem. J.*, **124**, 55–59.

41. Lisowska, E. and Duk, M. (1975), *Eur. J. Biochem.*, **54**, 469–474.

42. Furthmayr, H., Galardy, R.E., Schulte, T.H., Stryer, L. and Marchesi, V.T. (1976), (submitted).
43. Silverberg, M., Furthmayr, M. and Marchesi, V.T. (1976), *Biochemistry,* 15, 1448–1454.
44. Grefrath, S.P. and Reynolds, J.A. (1974), *Proc. natn. Acad. Sci. U.S.A.,* 71, 3913–3916.
45. Cotmore, S.F., Furthmayr, H. and Marchesi, V.T. (1976), (submitted).
46. Tanford, C. (1973), *The Hydrophobic Effect: Formation of Micelles and Biological Membranes,* Wiley, New York.
47. Braun, V. (1975), *Biochim. Biophys. Acta,* 415, 335–377.
48. Inouye, M. (1974), *Proc. natn. Acad. Sci. U.S.A.,* 71, 2396–2400.
49. Marvin, D.A. and Hohn, B. (1969), *Bact. Rev.,* 33, 172–209.
50. Asbeck, F., Beyreuther, K., Köhler, H., Von Wettstein, G. and Braunitzer, G. (1969), *Hoppe Seyler's Z. Physiol. Chem.,* 350, 1047–1066.
51. Wickner, W. (1976), *Proc. natn. Acad. Sci. U.S.A.,* 73, 1159–1163.
52. Nakashima, Y., Wiseman, R.L., Konigsberg, W. and Marvin, D.A. (1975), *Nature,* 253, 68–71.
53. Marvin, D.A. and Wachtel, E.J. (1975), *Nature,* 253, 19–23.
54. Nozaki, Y., Chamberlain, B.K., Webster, R.E. and Tanford, C. (1976), *Nature* 259, 335–337.
55. Lenard, J. and Compans, R.W. (1974), *Biochim. Biophys. Acta,* 344, 51–94.
56. Hughes, R.C. (1974), *Prog. Mol. Biol. Biophys.,* 26, 191–268.
57. Utermann, G. and Simons, K. (1974), *J. Mol. Biol.,* 85, 569–587.
58. Spatz, L. and Strittmatter, P. (1971), *Proc. natn. Acad. Sci. U.S.A.,* 68, 1042–104(
59. Wallach, D.F.H. and Zahler, H.P. (1966), *Proc. natn. Acad. Sci. U.S.A.,* 56, 1552–1559.
60. Vanderkooi, G. (1972), *Ann. N.Y. Acad. Sci.,* 195, 6–15.
61. Senior, A.E. (1973), *Biophys. Acta,* 301, 249–277.
62. Hudson, B.G., Wegener, L.J., Wingate, J.M. and Carraway, K.L. (1975), *Comp. Biochem. Physiol.,* 51B, 127–135.
63. Fujita, S. and Cleve, H. (1975), *Biochim. biophys. Acta,* 406, 206–213.
64. Hakomori, S. and Kobata, A. (1974), In: 'The Antigen' (Sela, M., Ed.), Vol. 2, pp. 79–140. Academic Press, New York.
65. Fujita, S. and Cleve, H. (1975), *Biochim. biophys. Acta,* 382, 172–180.
66. Lennarz, W.J. (1975), *Science,* 188, 986–991.
67. Lunney, J. and Ashwell, G. (1976), *Proc. natn. Acad. Sci. U.S.A.,* 73, 341–343.
68. Edidin, M. (1972), In: 'Membrane Research'. (Fox, C.F., Ed.), Academic Press, New York.

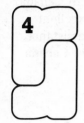

Specificity of Membrane Transport

M. SILVERMAN
Department of Medicine,
University of Toronto.

Acknowledgements

This work was supported by the Medical Research Council of Canada, MT 4590. The author is indebted to Mrs S. McGugan for excellent secretarial assistance and Miss R. Rutherford for technical assistance. M. Silverman is a scholar of the Medical Research Council of Canada.

4.1 BIOLOGIC SPECIFICITY – GENERAL CONSIDERATIONS

Prokaryotic and eukarytic cells control their internal milieu by regulating exchange of physical and chemical information to and from the external environment. In mammalian systems, this capability is vested in the functional and structural properties of the enveloping plasma membrane. The purpose of this chapter is to focus attention on a single aspect of membrane function – transmembrane transport. In particular we shall be emphasizing the specificity characteristics of transport mechanisms. To orient the discussion it is useful to make some general introductory remarks about biologic specificity before applying these concepts to transport phenomena.

From many observations it seems evident that the specificity of biologic reactions is one of the dramatic characteristics that distinguishes living systems from the world of the inanimate. The expression of this property is manifested at both macroscopic and microcospic levels and is fundamental to an understanding of biologic activity.

The term 'specificity' describes the ability to select a unique set of reactant structures from a larger population of nearly identical physico-chemical species. One of the remarkable properties of living organisms is that this exhibited selectivity is of such a high order. Perhaps the most familiar example is the stereospecific nature of enzymatic reactions in which only one enantiomer (d or l) is an appropriate substrate.

In the vast majority of cases, it turns out that the origin of the specificity of a biologic reaction is vested in particular protein structures. The processes of Transcription and Translation, operating according to the dictates of the 'Central Dogma' result in transformation of a linear set of triplet nucleotides into a polypeptide chain with a unique amino acid sequence. This primary structure determines the 3-dimensional potential energy reactant surface in a given protein – the so-called 'active site'. Thus the molecular basis of the specificity of biologic reactions reduces to considerations of macromolecular geometry.

In general, there appear to be two distinct structural features that determine the specificity of an enzyme–substrate interaction. The first

133

encompasses the ability of the enzyme to catalyze specific types of chemical bonds on the substrate. The second refers to the ability of enzymes to differentiate between certain functional groups on the substrate molecule and thereby position the substrate properly on the catalytic site. Considerations of membrane transport specificity belong in this second category.

Transport phenomena involve exchange of non-covalent bonds leading to transmembrane vectorial translocation of substrate. In contrast, enzymes catalyze transfer of covalent bonds. Despite this difference, the principles involved in enzyme—substrate specificity are equally applicable to solute transport processes.

The specificity of an enzymatic reaction is dependent upon a favourable binding energy between active site and specific substrate. The first hypothesis put forward to explain the specificity of enzymes was the 'template' theory of Fischer [1]. This proposal suggested that an exact fit was necessary in order for reactant groups on the substrate to be brought in contact with catalytic groups on the enzyme active site.

Along the same lines, in 1940, Pauling and Delbrück [2] introduced the concept of complementarity. The essence of their idea is that reacting molecular species fit in a lock and key type mechanism so as to bring the molecules as close together as possible. In this manner, positively charged groups are brought near to negatively charged groups, electric dipoles are brought into suitable orientation, and linearity of H-bonding groups is optimized. The specificity of enzyme—substrate interactions is therefore seen to depend upon promotion of bonding tendencies between chemical groups on the enzyme active site with chemical groups on the 'docking' substrate. Thus, forces such as Van der Waals and hydrogen bonding which decrease strongly with distance are maximally operative only when substrate is in the vicinity of the enzyme cleft (active site) and the dynamics is determined by the 'weak-bonding' tendencies of the interacting molecular species.

Koshland and his colleagues added a new dimension to our understanding of biologic specificity by introducing the induced fit theory of enzyme action [3]. This hypothesis is based on three postulates:
(1) that the active site can be modified by interaction with small molecules;
(2) that precise alignment of catalytic groups exists,
(3) that substrates cause conformational changes in the enzyme leading to correct alignments of catalytic groups whereas non-substrates do not. The novel feature in this proposal is the concept of enzyme (or active

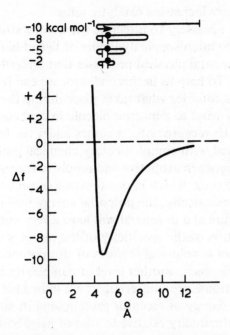

Fig. 4.1 Potential energy diagram for drug — receptor interaction. The curve has been calculated assuming an equilibrium distance between atomic centres of 5.0 Å and a free energy of interaction of -10 kcal mol^{-1}. The attractive forces have been divided equally among r^{-3} and r^{-6} forces, and the repulsive force has been assumed to vary as r^{-9}. Ordinate Δf in kcal mol^{-1}. The abscissa is the separation between the atomic centres (Å). In the upper part of the diagram there is shown in a schematic simplified way the behaviour of molecules initially at the equilibrium distance. The arrows indicate their behaviour on acquiring kinetic energy as indicated in the direction normal to the receptor surface. From [5].

site) flexibility.

When combined with complementary structural features on reactant and enzyme, a substrate-induced conformational state enables a portion of the binding energy to be utilized as the driving force for catalysis. In the case of an induced fit, most of this binding energy is consumed to force a change from inactive to active enzyme conformational state. As discussed by Jencks [4], binding energy may also be utilized more directly to influence the rate of catalysis by destabilization mechanisms, non-productive binding and entropic losses as well as via the indirect substracte induced changes. All of these mechanisms acting singly or in combination, enable enzymes to bring about a microscopic state of higher order, thereby increasing catalytic rates.

Ultimately, it is necessary to integrate the unique structural features of proteins with the microscopic dynamics of ligand binding in order to explore the fundamental physical processes that underlie the kinetics of biologic reactions. To help us in this endeavor we can follow the lead of Burgen [5]. Let us consider what takes place during the binding of a small ligand (e.g. a drug) to a macromolecule in aqueous solution. As the ligand approaches its receptor site, it comes under the influence of a force field associated with complementary chemical groups on the active site. The balance between attractive and repulsive forces yields a potential energy well for the drug, which in one dimension can be represented as in Fig. 4.1. More realistically, the potential energy well is a three-dimensional structure and in general will have a very complex surface. A precise representation of the specificity of the receptor-drug interaction is given by the exact topological features of this potential energy well.

In the membrane phase, another level of complexity is introduced. In this medium, functional protein receptors are imbedded in a non-aqueous phase and the free energy of receptor protein (and its subunits) is determined by its amphipathicity relative to that of neighbouring phospholipids. As a result, it is the thermodynamically preferred conformational state of a given membrane protein that influences its 'specificity' for an interacting ligand. We must also remember that the dynamics of receptor − drug interactions are drastically altered by other factors such as the screening effect of solvent (water) molecules and occupancy of active sites by ions or other bound chemical species [5].

In summary, although the scheme shown in Fig. 4.1 is oversimplified, it provides us with a useful way of visualizing the determinants of the specificity of interaction between a small ligand and a biologically active macromolecule. We now turn to the special case of membrane transport processes where ligand − receptor interactions involve physical translocation of solute across the ∼ 100 Å width of the plasma membrane.

4.2 CURRENT CONCEPTS OF MEMBRANE TRANSPORT

The molecular basis of transmembrane transport is vested in a specialized group of membrane-bound proteins which are called 'transport receptors' or 'carriers'. In order to understand their mechanism of operation, we first require precise information about how they are integrated into the structure of the plasma membrane.

Modern concepts of membrane structure are contained in the fluid mosaic model [6]. This scheme provides a useful framework for a description of the dynamic behavior of membrane proteins in general, and transport proteins in particular.

According to this model, membrane lipid is in the form of phospholipid organized principally in the thermodynamic lowest free energy state as a bimolecular leaflet with charged polar headgroups interacting with the external aqueous media and hydrocarbon alkyl side chains making up the hydrophobic interior. At physiologic temperatures the phospholipid is primarily in the fluid state with considerable lateral diffusive mobility. Lateral phase separations, exist which effectively create more 'frozen' patches [7, 8]. These limit mobility of the fatty acid alkyl side chains in localized regions of the membrane interior and secondarily influence the function of membrane protein.

Some membrane proteins have a significant hydrophilic interface with the exterior aqueous medium and minimal hydrophobic interaction [9]. These superficially located molecules can be leached off the surface by means of relatively gentle treatments that disrupt ionic bonding forces between lipid headgroups and the hydrophilic amino acid side chains of the protein. This generally leaves the remainder of the plasma membrane structure intact. Other membrane proteins have a more initimate association with neighbouring lipids. These are the so-called 'core' or 'integral' membrane proteins [9]. They are imbedded deeply in the membrane matrix and solubilization procedures (i.e. using nonionic detergents) which break hydrophobic interactions not only disperse these protein constituents but also completely disrupt the membrane structure. Fig. 4.2 shows a schematic representation of the Singer-Nicholson Model.

An essential feature of the organization of membrane phospholipids and proteins is their asymmetrical distribution along the axis perpendicular to the membrane. Thus $80 - 85\%$ of sphingomyelin and $65 - 70\%$ of lecithin is located on the outer surface of the erythrocyte membrane while phosphatidylserine and phosphatidylethanolamine have reactive amino groups accessible only from the cytoplasmic (inner) membrane surface [9]. Similarly, membrane glycoproteins are accessible only to the exterior surface while other more intrinsic proteins are buried and exposed at the inside surface.

Experimental evidence from a variety of different mammalian cells strongly supports the concept that all carrier proteins are intimately associated with membrane lipid and are therefore examples of 'core

EXTERIOR

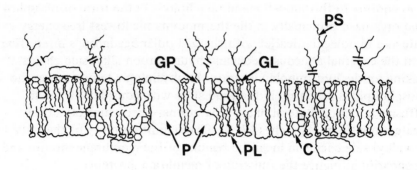

INTERIOR

Fig. 4.2 Classical diagram of a biomembrane. Cholesterol (C), proteins (P), glycoproteins (GP) and glycolipids (GL) are submerged to various depths in a bilayer of phospholipids (PL). PS, polysaccharide chains. ‖, deleted midportion of very long polysaccharide chain. From [27].

proteins'. For example, the glucose transporter in the brush border of the proximal tubule in dog kidney lies spatially deep to the more superficially located aminopeptidases and disaccharidases [10, 11, 12]. Thus according to Singer's classification described above, the glucose transporter is an example of an 'integral' membrane protein.

We now come to the question of whether transport systems operate as transmembrane channels or as mobile carriers. Examples of the former are gramicidin A and alemethicin [13]. These channels operate even when the membrane is frozen below the transition temperature of the phospholipid [14]. The prototypes of mobile carriers are valinomycin or nigericin which can diffuse across the membrane only if it is in a fluid state. Singer [9] has presented thermodynamic arguments aginst any transport model requiring rotational or translational excursion of carrier. In addition, there is recent experimental information which favors the view that transport systems operate as transmembrane channels.

(Na^+ and K^+) ATPase is an enzyme complex composed of two subunits –

a large chain (mol, wt. 130 000) and a small chain (mol. wt. 40 000) [15]. It is this enzyme which is responsible for the coupled active transport of Na^+ and K^+ in opposite directions across the plasma membrane of animal cells [16]. The small chain of (Na^+ and K^+) ATPase is a sialoglycoprotein exposed at the external surface of the cell [15] while the large chain spans the membrane [17]. Anti-large-chain antibody binds to specific antigenic sites which are on the large chain of the enzyme and are located on the cytoplasmic side of the plasma membrane [17]. Anti-holoenzyme antibody recognizes both small and large chains and binds to both sides of the plasma membrane [18]. Ouabain, which inhibits (Na^+ and K^+) ATPase activity and blocks Na^+ transport, binds to the large chain at the exterior surface [17, 19]. Thus the enzyme responsible for trans-membrane pumping of Na^+ and K^+ is most readily visualized as forming some type of transmembrane channel, spanning the width of the bilayer.

Henderson and Unwin have recently reported the resolution of the purple membrane of *Halobacterium halobium* at the 7 Å level by electron microscopy [20]. Seventy-five percent of the total mass of this membrane is made up of identical proteins of molecular weight 26 000. Moreover these proteins function *in vivo* as light-driven H^+ ion pumps [21]. Therefore this can be taken as an example of a specific transport protein system. It turns out that this H^+ ion transporting system contains seven closely packed α-helical polypeptide segments. The overall dimension is $25 \times 35 \times 45$ Å with the largest dimension perpendicular to the plane of the membrane and parallel to the helices. Fig. 4.3 reproduces a model of the bacterial rhodopsin taken from reference 20. The inner ring of nine helices has a circumference of approximately $80 - 90$ Å in diameter. Moreover, the pore is filled with lipid molecules arrayed in the bilayer configuration, and not with solvent (i.e. H_2O).

The importance of the work on the purple membrane is that it is the first accurate description of the structure of a transporting membrane protein within its natural milieu, i.e. the plasma membrane. Racker and Hinkle [22] have incorporated bacterial rhodopsin into vesicles made of dimyristoyl phosphatidylcholine. They showed that the reconstituted H^+ pump was operative at temperatures where the phospholipid was frozen.

On the basis of the foregoing definitive experimental evidence it is now generally accepted that transport proteins span the entire membrane thickness and have hydrophilic residues exposed at opposing surfaces (extracellular vs intracellular). For some bacterial transport systems, there is probably an additional binding protein located geometrically

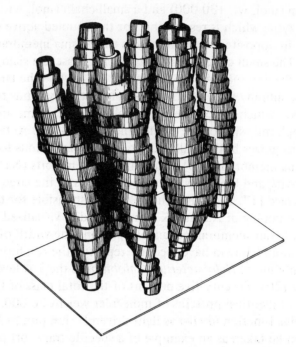

Fig. 4.3 A reconstruction of the proton-transporting transmembrane protein from *Halobacterium halobrium* (see text for details). From [20].

superficial to the true transporter.

Now that we have some rough idea as to the relationship between carriers and the lipid constituents we turn to the question of the distribution of these transporters in the plane of the membrane. As in the case of other receptors, 'carrier' molecules seem to occur at very low densities within the plasma membrane. To illustrate this point, there are about 12 picomoles of glucose transporter per mg membrane protein [23] in the brush border of the dog kidney proximal tubule. For a single red cell, 1.3×10^{-12} g dry weight [24] is distributed over a surface area of 135 μm^2 [25]. Assuming 50% is protein we can calculate that there is $\sim 0.65 \times 10^{-12}$ g protein/100 μm^2 or $\sim$ 4800 molecules of glucose transporter (phlorizin receptor) per 100 μm^2 of brush border membrane (i.e. $\sim 2 \times 10^5$ per proximal tubular cell).

Multiple transport systems co-exist in the same cell. Each is designed to translocate a different class of chemical compound such as ions, sugars and amino acids. A reasonable question to ask is how much membrane protein is devoted to carrying out these different transport processes? Lumping all of the various transported species together (including both non-electrolytes and ions) it is unlikely that there are more than 10—20 different carriers per cell. In dog kidney brush border, there appear to be more than 30 different proteins as identified on polyacrylamide gel electrophoresis (M. Silverman, unpublished results). If the molecular weight of the glucose carrier is assumed to be $\sim 10^5$ daltons, then this transporter represents something of the order of 0.5% of the total membrane protein. Therefore the sum of all different transporters in a given mammalian cell is probably no more than 5—10% of the total membrane protein. The remaining proteins are specialized for such functions as cell recognition (e.g. HLA, lectin receptors), metabolic regulation (hormone receptors and adenylcyclase) and enzymatic catalysis (e.g. carbonic anhydrase, alkaline phosphatase and aminopeptidase).

4.3 MOLECULAR MECHANISM OF ACTION OF TRANSPORT PROTEINS

We are now in a position to begin to speculate as to the actual chemical events that might be taking place at the atomic and molecular level during the transmembrane trasnport process. If carrier protein(s) extend across the membrane width (see Fig. 4.4), then the specificity of

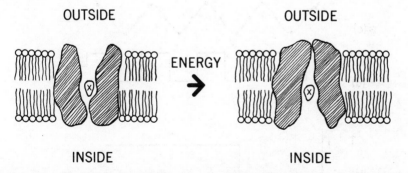

Fig. 4.4 A schematic mechanism for the translocation event in active transport. A specific site for a hydrophilic ligand X exists on the surface of a pore formed by a particular subunit aggregate; the aggregate might actually be tetrameric, like hemoglobin. Some energy-yielding process is then converted into a quaternary rearrangement of the subunits, which translocates the binding site and X from one side of the membrane to the other. Reversal of the protein change restores the subunit aggregate to its initial state. From [9].

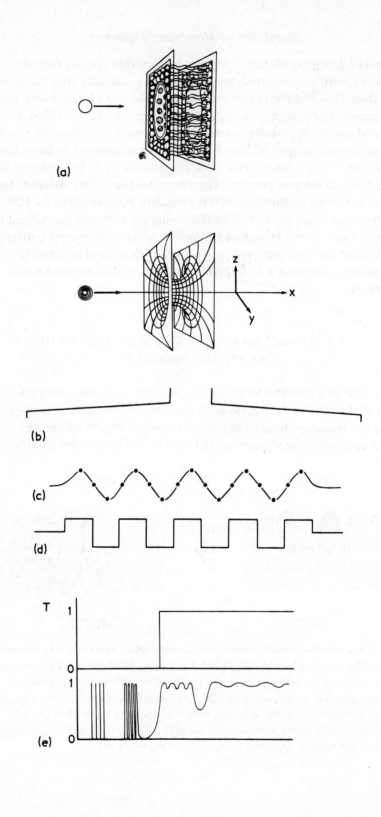

(a)

(b)

(c)

(d)

(e)

'docking' will be determined by certain amino acid side chains projecting into the pore entrance at either surface. The likelihood of bond formation between substrate and 'receptor' will be a function of steric considerations and the hydrophobicity of the amino acid side chains. The latter will in turn be a function of the local microenvironment, e.g. adjacent phospholipid head groups and bound water molecules. Conformational changes in the transmembrane channels could also be accomplished by binding of ligands to non-transport sites. This might be a specific mechanism for regulating passage of molecules and in particular accomplishing unidirectional pore traffic (Fig. 4.4).

What about the physics of the actual binding event and transmembrane translocation process. In particular, we must ask ourselves if macroscopic continuum concepts about transport will be useful within the microscopic realms of receptor sites ($\lesssim 10$ Å) and membrane interiors ($\lesssim 100$ Å).

In previous work from our laboratory [26, 27] the quantum structure introduced into molecule — membrane interactions was assessed for potential energy gradients that approximate a membrane barrier. Classically, when the energy of an incident molecule is less than that of the potential energy barrier, the molecule is reflected and hence impermeant. Exact molecular position, velocity and orientation can in principle be predicted with certainty. By contrast, in quantum mechanical considerations, the molecule is characterized by a wave packet of finite width in the space of its conformational co-ordinates. During interaction with non-zero membrane potential gradients, the packet structure changes non-classically. Trajectory and conformation are not certain, and this branching spectrum of possibilities enriches the molecular dynamics.

Recently, Lumsden [28] has developed classical and quantum flux equations to describe the transmission of small substrates such as protons, water and Na^+ ions through potential energy structures which model (a) lipid bilayers and (b) pore structures such as may exist for gramicidin

Fig. 4.5 (a) Schematic representation of a molecule incident on a transmembrane pore. (b) Schematic representation of the potential energy surface characterizing the molecule — pore interaction. (c) The potential energy structure along the pore axis — x. (d) An idealized representation of the one-dimensional potential energy structure. (e) The transmission T for two cases. The upper diagram shows the result for the classical case where transmission necessitates that the energy of the incident molecule be greater than the height of the potential energy barrier; the lower diagram indicates the quantum transmission pattern which appears following interaction with a multiple potential barrier system showing transmission resonances below the classical turning point.

(an ion-selective antibiotic). These equations permit a precise assessment of the role of purely microphysical processes *per se* in the net model transport behavior. As might be expected, quantum corrections to the classical fluxes are negligible for the bilayer models. The quantum processes can, however, take on significant magnitude when the bilayer potential barrier is modified to assume the characteristics of a hydrophilic, pore-like structure.

The extrapolation of this type of theoretical modelling to the H^+-conducting properties of the purple membrane is an exciting prospect. In general it may turn out that all transport proteins span the bilayer as shown schematically in Fig. 4.4. Then for transmembrane passage of a hydrophilic solute, it seems reasonable that there must exist some kind of channel through the membrane lined by hydrophilic amino acid residues. The channel might be formed by a configuration of several α-helical subunits. The important feature of this type of model is that a transported solute wave packet upon entering such a channel will 'see' a multiple potential energy barrier structure. This fine structure forms a secondary lattice convolution with the superlattice unit cell formed by the bilayer. We might then consider proton transmission through membrane channels of the type shown in Fig. 4.5 as a quantum mechanical process with the hydrogen bonds along the α-helix perhaps forming defined potential 'wells'.

Such considerations can be extended to larger molecular species such as sugars. For these substrates, the necessary quantization of internal modes (vibrational, rotational, translational) must be taken into account. The combination of spatial form with a rich spectrum of discrete internal modes is expected to confer unique substrate selectivity predictions to the quantum statistical transport models. In this way we hope to gain deeper insights into the biologic relevance of the 'specificity' of transport processes.

4.4 SUGAR TRANSPORT ACROSS THE PROXIMAL TUBULE OF DOG KIDNEY

4.4.1 General properties

In parallel with the extensive cell differentiation characteristic of mammalian systems, there has occurred differentiation of transport systems for a given molecular species. For example, comparison of the d-glucose transport mechanism in erythrocytes, muscle, and kidney tissue exposes marked differences in terms of Na^+ dependence, substrate specificity, insulin-dependence and kinetic behavior. This high degree of transport specialization is reflected in the different functional capabilities of the individual cell. The discrete nature of certain inborn errors of metabolism that can be traced to defects in membrane transport is another manifestation of this specialization process. As an illustration, inherited renal glucosuria is a consequence of an alteration in the renal reabsorptive mechanism for d-glucose. Evidence has been presented in man, that absent intestinal glucose transport is shared by a partial defect in renal glucose transport [29]. This and other observations indicate the presence of two genetically distinct renal glucose transport sites in man, one shared with the gut and one confined to the kidney [30]. Thus, even for an homorphous substrate species such as pyranoses, the specificity of transmembrane transport is carefully integrated with the global metabolic functions of the individual cell and each transport system appears to be under separate genetic control.

In epithelial structures, there is definite polarization of membrane structural and functional properties. In particular, there exist different transport systems for the same substrate at opposing cell surfaces (luminal vs antiluminal). This situation provides a unique model from which to gain insight into some of the biologic implications of transport receptor specificity. For this reason we shall now concentrate on the specificity characteristics associated with sugar uptake in a particular tissue — the kidney proximal tubule.

The transit of a molecule of glucose across the tubular epithelium from urine to blood involves in sequence three distinct transport processes: —
(1) transport across the brush border (luminal membrane);
(2) diffusion through the intracellular space;
(3) transport across the basal or lateral cell surface (antiluminal membrane).
It is essential to note that the reabsorption process involves no breakdown or rearrangement of the carbon skeleton during the transepithelial passage [31].

Relatively little is known about the glucose transport process at the antiluminal surface of the proximal tubular cell in contrast to the more extensively characterized glucose transporter at the brush border membrane. This is true not only for the kidney cell but also in the intestine and in every other example of a polarized epithelium.

The Na$^+$-coupled sugar transport processes present in mammalian epithelium could represent differentiation from the primary active transport processes which exist in more primitive organisms. In the same way sequential *d*-glucose transport across the luminal and antiluminal membranes of tubular epithelia may have evolved from that of a non-polarized cell. If this were the case, then glucose transport at the two opposing plasma membrane surfaces may be functionally coupled into an integrated operational unit. This is only one of many interesting problems posed by transporting epithelia. In order to investigate such systems experimentally it is necessary to study *separately* the transport of substrates such as *d*-glucose across the individual surfaces of the proximal tubular epithelium. This has been made possible *in vivo* by adaptation of the multiple indicator dilution technique [32] and *in vitro* by the isolation and separation of membrane vesicles derived from opposing surfaces of the tubular cell [33].

4.4.2 Carbohydrate chemistry

In a series of *in vivo* studies we have investigated the specificity of sugar transport in dog kidney [34, 35]. The rationale for our studies has been to make use of known structural and conformational properties of sugars in aqueous solution to deduce the chemical and steric deter-minants which underly their interaction with the surfaces of the proximal tubule. To better appreciate the specificity characteristics for this particular experimental system it is worth expanding briefly on some general aspects of sugar − water and sugar − lipid interaction.

The conformation of monosaccharides in aqueous solution has been extensively studied by Reeves [36] using cupraamonium complexes. He proposed that there are two rigid chair forms designated 1C and Cl and an infinite number of flexible 'boat forms' which may be described in terms of six specific conformers whose interconversion involves little ring strain. In modern nomenclature the C1 and 1C conformation are referred to as $^4C_1$ and $^1C_4$ respectively. The letter C refers to a *chair* conformation. The pyranose ring is numbered as in Fig. 4.6. The shaded area defines a reference plane containing four ring atoms. Ring atoms which lie above the reference plane (numbered clockwise from above) are written as superscripts, while ring atoms which lie below the refer-ence plane are written as subscripts. Interconversion of the rigid chair, involves considerable deformation and rotation of the ring valences. Most of the aldopyranosides exist predominantly in the chair forms.

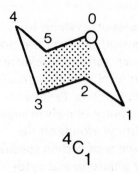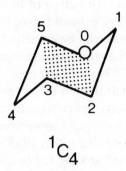

Fig. 4.6 Schematic representation of conformational representation of aldohexoses. For explanation see text.

Equatorial groups on adjacent carbon atoms are directed within 30° of the plane of the ring but do not eclipse each other, while axial groups are nearly perpendicular to the plane of the ring. An axial position creates instability and the chair conformation most preferred is that in which the majority of the bulky or polar groups are oriented equatorially.

The monosaccharides can be assigned instability ratings which depend upon the presence of axial groups and the relative weighting factors attached to them. The greater the instability rating the greater the conformational variations, and presumably, the weaker the solute − water interactions. The importance of these conformational variations in determining the partioning of monosaccharides between different phases was borne out by the work of Marsden [37]. He found that a low instability rating was associated with exclusion from a gel phase. In other words, the greater the sugar − water interaction, the less the partitioning between solvent and gel.

Since these data suggest a rather strong sugar − water interaction, it might be expected that sugars would tend to be excluded from biological membranes. However, sugars are rapidly taken up by cells, and this implies that some special catalytic process is required to break the intermolecular forces holding sugars (in particular glucose) in aqueous solution.

Is there any specific sugar − lipid interaction? The ether − water partition coefficients of monosaccharides are extremely low. For example, for *d*-glucose, Wright and Prather [38] quote a value of 4.5×10^{-6}, compared to 1, 6-hexanediol (also a six-carbon compound) which has a value of 1.2×10^{-1}. There is other experimental work [39] indicating that the basis of sugar − lecithin interaction is consistent with the formation of hydrogen bonds.

In very general terms, it appears that sugar transport systems in mammals can be classified into two broad categories, although it is not clear that these two categories of sugar transporters reflect basic differences in operational mechanism at the molecular level. The first group, of which the red cell is the best-studied example, exhibits: (a) low specificity for sugar substrates, i.e. the transportability of a given sugar seems to depend more upon the number of hydroxyl groups on the pyranose ring rather than on the presence or absence of certain specific hydroxyl groups; (b) sensitivity to inhibition by phloretin and cytochalasin B with relative insensitivity to phlorizin; (c) lack of dependence on sodium (extracellular or transmembrane) concentration and (d) facilitated diffusion kinetics (asymptotic steady state ratio of intracellular to extracellular glucose concentration equal to one). The second class of sugar transport systems, of which the prototypes are the kidney and intestine, is characterized by (a) high specificity for sugar substrate i.e. critical dependence on certain hydroxyl groups; (b) competitive inhibition by phlorizin but not phloretin; (c) sodium dependence and (d) active transport kinetics (i.e. transport against electrochemical potential gradients resulting in intracellular/extracellular glucose concentration ratios greater than one).

4.4.3 Sugar transport systems in the proximal tubular cell

Now let us return to the results for the dog kidney proximal tubule. Competitive inhibition studies reveal four apparently distinct sugar transport receptors as shown diagramatically in Fig. 4.7 [23, 32, 34, 35, 40]. Three of these are localized at the brush border:
(1) the G or glucose transporter, which is shared by *d*-glucose, α (β)-methyl-*d*-glucopyranoside, *d*-galactose, 2-deoxy-*d*-glucose, *d*-fructose and myoinositol.
(2) the M or mannose transporter (also shared by *d*-fructose);
(3) the Myo for myoinositol transporter.

At the antiluminal, or basal surface, the present evidence suggests that there is only one transporter (called G′ in Fig. 4.7) in dog kidney (M. Silverman, *Am. J. Physiol.,* in press), which is shared by *d*-glucose and at least nine other sugar substrates (*d*-galactose, *d*-mannose, 3-*O*-methyl-*d*-glucose, 6-deoxy-*d*-galactose, *d*-xylose, *d*-talose, *l*-arabinose, myoinositol and *d*-fructose).

Glucose transport at the luminal surface (G transporter) is exquisitely

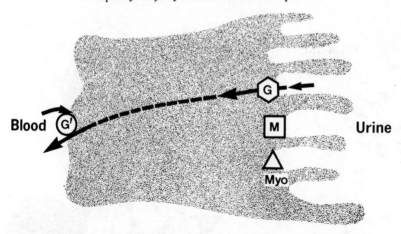

Fig. 4.7 Schematic representation of proximal tubular cell indicating
3 sugar transport systems at the brush border membrane, G for the glucose
transporter, M for mannose and Myo for myoinositol (refer to text). At
the antiluminal membrane we have a different glucose transporter, G′ shared
by 9 other substrates (see text). The implication of this diagram is that a
molecule of *d*-glucose which is reabsorbed across the brush border of the
proximal tubule diffuses across the cytoplasm and exits via the G′ glucose
transporter in the antiluminal membrane.

sensitive to the competitive inhibitor phlorizin [32] and is dependent
upon transmembrane Na^+ concentration (lumen → cell) [33]. Conversely,
transport via the G′ transporter at the antiluminal membrane is
~ 1000 x less sensitive to inhibition by phlorizin and is not dependent
on a Na^+ concentration gradient [32, 33].

Much of our knowledge about the characteristics of the glucose
transporter in different tissues and the kidney in particular is derived
from studies of the interaction of the drug phlorizin.

Fig. 4.8 shows a space-filling molecular model of phlorizin along with
its aglycone phloretin. The latter has at least 100 times less potency in
inhibiting glucose transport at the brush border membrane.

4.4.4 Mechanism of phlorizin binding

Phlorizin has about 1000 x greater affinity for the glucose transporter
at the brush border membrane than does glucose itself under *in vivo*
conditions. By correlating the structure of phlorizin analogues with
their inhibitory potency on glucose transport in the kidney, it has been

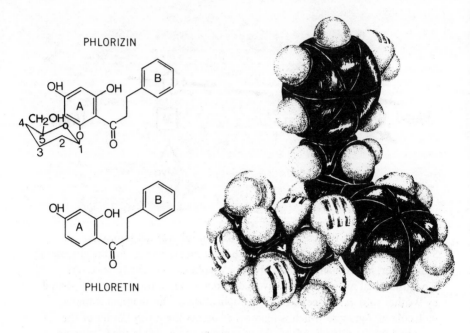

Fig. 4.8 Schematic representation of phlorizin and its aglycone phloretin on
the left with a perspective drawing of a particular conformation of the
phlorizin molecule shown on the right. It should be emphasized that there
are many possible stable conformational states. The aromatic rings make up
a hydrophobic portion whereas the glycoside moiety represents the
hydrophilic moiety of this amphipathic drug.

possible to deduce that the phlorizin molecule has several points of
attachment to the glucose transporter receptor [41];
(1) via the glucoside moiety, and
(2) via hydroxyl groups on the aromatic A and B rings (see Fig. 4.8).
The major determinant of binding specificity resides in the glucoside
moiety.

Diedrich [41] has suggested a possible active conformation for
phlorizin. The essence of his proposal is that there is an intramolecular
hydrogen bond between the carbonyl oxygen and the hydroxyl group at
the carbon 6 position of the glycoside moiety. This hydrogen bond fixes
the glucoside moiety in the plane of ring A. The OH group at position 4
in the B ring is visualized as binding out of the plane of the glucoside
moiety and at a distance of $\sim 12 - 15$ Å from the oxygen at position 4
of the pyranoside. Because our studies [34, 35] suggest that the hydroxyl

group at carbon 6 is crucial for the successful transport of pyranosides at the brush border, we have argued that the conformation of phlorizin proposed by Diedrich may exist in aqueous solution but not when bound to the glucose transporter.

The recent proposal of a 'zipper' binding mechanism for extended polypeptides to surface receptors provides an obvious alternative model to explain the observed high affinity of phlorizin for the glucose trans-porter [42]. The binding of different functional hydroxyl groups on the phlorizin molecule to the glucose transporter might occur sequentially rather than simultaneously. For example, initially there might occur nonspecific hydrogen bond formation at or near the glucose transporter via the OH groups in rings A and B. This initial nonspecific interaction would reduce the free energy of binding of the glucoside moiety thereby enhancing the observed total binding affinity and offering a distinct kinetic advantage for the phlorizin molecule compared to d-glucose with respect to its receptor interaction. Since the phlorizin molecule possesses both a hydrophobic moiety (i.e. phloretin) and a hydrophilic moiety (glucoside), its behavior could be viewed as an amphipathic structure which interacts with hydrophobic regions (i.e. either hydrophobic residues on the transporter or adjacent lipids) as well as the exteriorized hydrophilic portion of the receptor.

4.4.5 Specificity characteristics of glucose transporters (G and G′) at opposing surfaces of the proximal tubule

Now let us examine the results of the specificity characteristics of the G and G′ transporters (see Fig. 4.7). These have been obtained by systematically comparing the renal tubular interaction of an homologous series of 23 different pyranose and pyranose derivatives. In this way, we have been able to deduce the chemical and steric determinants of pyranose interaction with the sugar transporter at each of the two surfaces of the proximal tubule.

Antiluminal membrane
The minimal specificity criteria for the d-glucose transporter (i.e. G′ transporter) at the antiluminal membrane include:
(1) the necessity for an hydroxyl (OH) group being present at the carbon 1 and carbon 2 position of a cyclohexane ring in the $^1C_4$ or $^4C_1$ chair conformation, oriented as in the d-glucose configuration;
(2) although there is no requirement for an OH group at the carbon 3 or

6 position, these OH groups, if present, must be equatorial.

Luminal membrane

The minimal specificity characteristics governing pyranose interaction with the glucose transporter at the brush border membrane (G transporter) consist of:

(1) a pyranose ring in the $^1C_4$ or $^4C_1$ chair conformation with hydroxyl substituents at C3 and C6 oriented as in the configuration of *d*-glucose, i.e. equatorial;

(2) if present, the OH group at the carbon 2 position must also be equatorial.

From these considerations it is apparent that the steric arrangement of certain functional OH groups on the pyranose ring is a more important determinant of pyranose-transporter interaction than is the type of chair conformation, $^1C_4$ or $^4C_1$. To date our investigations have revealed no other single chemical or physical parameter in the pyranose family that correlates with the observed pattern of specificity of transport.

Up to now we have described the minimal requirements governing sugar interaction with the glucose transporter. It is not surprising that glucose is the optimal substrate for both the G and G' carrier systems.

From our experiments, we have deduced that the glucose molecule interacts with its brush border transport receptor (G) through 5 different points of contact on the pyranose ring: via OH groups at positions C2, C3, C4, C6 and oxygen of the pyranose ring. Both minimal and optimal specificity determinants are shown schematically in Fig. 4.9. Energetically, the most reasonable hypothesis would be to postulate non-covalent bond formation between the transported glucose molecule and its carrier at each of the five points of contact shown in Fig. 4.9.

Referring back to the scheme depicted in Fig. 4.1, we therefore visualize a glucose molecule 'homing in' on its G transporter as shown in Fig. 4.9. Presumably the hydrophilic amino acid side chains at the entrance of the transport molecule form a complementary weak bonding pattern with the OH groups on the pyranoside. Important determinants of the specificity of the sugar — transporter interaction might also be manifested in such a situation by a steering group(s) on the glucose molecule providing critical 'manoeuvring' as the sugar 'docks' in the receptor force field.

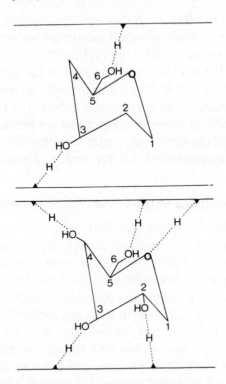

Fig. 4.9 The minimal (upper) and optimal (lower) interacting sites between a pyranoside and the brush border membrane (from the luminal surface). A non-covalent interaction (H bond) is tentatively assigned as indicated.

4.4.6 Role of Na⁺ ion in specificity of sugar transport

We have already alluded to a special role for Na⁺ in sugar transport in the kidney and other tissues. Glucose transport at the brush border is electrogenic and involves co-transport of a Na⁺ ion without co-movement of an anion or counter movement of a cation via the same carrier [33, 43]. Moreover the presence of a Na⁺ gradient (lumen → cytoplasm) provides energy for the active transport of glucose. By comparison, glucose transport at the antiluminal membrane is not dependent on the presence of Na⁺ [33].

Kolinska [44] has evidence in rabbit kidney cortex slices that Na⁺ somehow facilitates the rate of translocation of certain sugars rather than altering the apparent affinity constant for the transporter.

There is other data that suggest the existence of two glucose carrier states in the brush border, one Na^+-dependent and one Na^+-independent [45]. Along the same lines, results from our laboratory are beginning to implicate the Na^+ ion in a carrier interaction at the brush border that specifically alters the binding of sugar ligand at the carbon 2 position. Thus it still remains possible that the apparently different mannose (M) and glucose (G) transporters in the brush border (see Fig. 4.7) are not genetically independent transporters but instead reflect different specificity states of the same transporter — each state being controlled by the local microenvironment (i.e. Na^+ and H_2O concentrations).

4.4.7 Functional coupling of the G and G′ transporters

Consideration of the foregoing specificity data indicates that the brush border and antiluminal membrane in dog kidney proximal tubule appear to be acting as two 'selectivity' barriers in series in which the transport specificity criteria are complementary to one another. Thus the ability to penetrate the luminal membrane depends crucially on functional hydroxyl groups at C3 and C6 with less emphasis on interactions at the C1 position of the pyranose ring. Conversely, there is critical dependence on C1 and C2 hydroxyl interaction with the glucose transporter at the opposite side of the cell. Thus, glucose, which has all of the necessary functional equatorial hydroxyl groups, successfully traverses both barriers in sequence. In contrast, α-methyl-d-glucopyranoside is reabsorbed across the brush border with the same affinity as d-glucose but because it lacks an hydroxyl group at the C1 position, its reabsorption into the blood is negligible compared to d-glucose, over similar time scales of observation [46].

Recently, Barnett *et al.* [47] have presented detailed evidence for two asymmetric conformational states in the human red cell sugar transport system. Their findings suggest that sugar binds to the transport system so that C4 and C6 are in contact with the solvent on the outside of the membrane and C1 is in contact with solvent on the inner (cytoplasmic) surface. Following the data given in [47], a schematic representation of the specificity of substrate interaction with the glucose transporter in the human red cell is shown in Fig. 4.10.

Cytochalasin B is a potent inhibitor of glucose transport in the human erythrocyte. Taylor and Gagneja [48] demonstrated that the kinetics of inhibition of glucose exit from preloaded cells is competitive when cytochalasin B is present in the external medium. The K_i of inhibition was similar to the high affinity dissociation constant $\simeq 10^{-7}$ M.

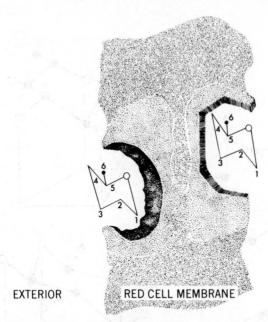

EXTERIOR   RED CELL MEMBRANE   INTERIOR

Fig. 4.10 Schematic representation illustrating the asymmetrical specificity characteristics of the red cell sugar transporter. The transporter is a protein spanning the width of the membrane, no attempt is made to define the interior, i.e. more hydrophobic portion of the transport system. Only the surface docking sites for pyranoses are shown. According to the specificity criteria from [47] the C4 and C6 hydroxyl groups are more intimately associated with the receptor site at the interior face of the membrane. At the exterior surface it is the hydroxyl group at the carbon 1 position which is most intimately associated with the receptor site.

These authors examined a molecular model of cytochalasin B (Fig. 4.11) and demonstrated that four of the inter-oxygen distances in cytochalasin B (as indicated in Fig. 4.11) are identical and superimposable on those corresponding to site A, B, C, and D of β-d-glucopyranose in the $^4C_1$ conformation (Fig. 4.12). These are precisely the OH sites implicated in the hydrogen bonding of d-glucose to its carrier protein in the red cell membrane [49].

Making use of the specificity data on sugar interactions with both surfaces of the proximal tubule, we can illustrate the asymmetries of transepithelial glucose transport as shown in Fig. 4.13. In the case of the dog kidney, our data indicate that sugars which share the glucose transport pathway bind to the transport system so that C1 is in contact with

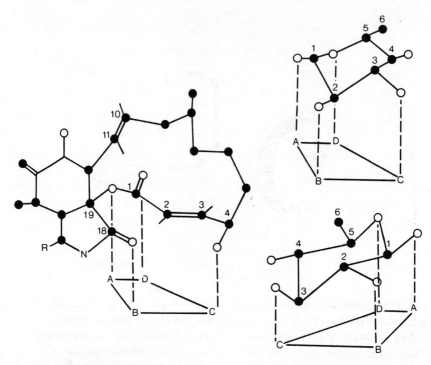

Fig. 4.11 Model of cytochalasin B.
Dashed lines represent hydrogen
bonds, A, B, C, and D represent
protein receptor sites.

Fig. 4.12 Upper diagram shows
model of $^4C_1$ conformation of
β-d-glucopyranose and the lower
one shows the inverted model.
A, B, C and D represent protein
receptor sites corresponding to
those shown in Fig. 4.11.

the solvent on both the exterior and cytoplasmic surfaces of the brush
border membrane, while at the antiluminal membrane sugar binding to
its transporter occurs so that the C4 and C6 portion of the pyranoside
molecule is in contact with the solvent at exterior and cytoplasmic
surfaces.

The erythrocyte plasma membrane is non-polarized in structure and
function in contrast to the renal proximal tubular cell. Yet it is of
interest to compare the sugar transport specificities in the two cell
types. As indicated above, the steps involved in sequential exit and
re-entry across the red cell membrane (Fig. 4.10) and that of sugar

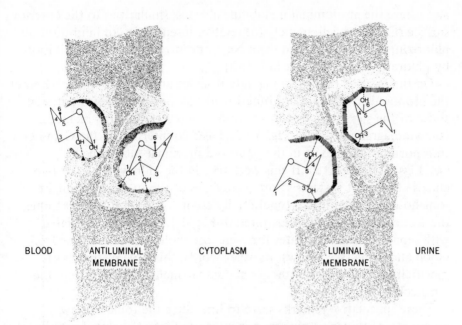

BLOOD ANTILUMINAL CYTOPLASM LUMINAL URINE
MEMBRANE MEMBRANE

Fig. 4.13 Schematic representation showing the specificity of sugar interactions with the luminal and antiluminal membranes of the proximal tubular cell in dog kidney. This diagram is meant to be compared with Fig. 4.10, showing sugar interactions with the red cell membrane. The interaction of pyranoses with the opposing cell membranes is derived from the specificity data in the text. Although not specifically discussed in the text, there is evidence from [32] and [34] and also from M. Silverman and L. Huang (*Am. J. Physiol.*, in press) and M. Silverman (*Am. J. Physiol.*, in press) that the specificities assigned to the cytoplasmic surface of the luminal and antiluminal membranes are the same as those assigned to the exterior surfaces of their opposing membranes.

reabsorption from urine to blood across the proximal tubular epithelium (Fig. 4.13) are almost identical. Each can be visualized as binding of (a) first the C4 and C6 portion of the pyranose ring followed by (b) the C1 portion of the pyranose ring. Thus the asymmetry of the glucose transport system across the width of the red cell membrane seems to correspond to the asymmetry of the G and G′ transporters localized at the two surfaces of the renal proximal tubular cell. This correspondence may be entirely fortuitous. However, the intestine seems to be another example of a polarized epithelium where the basal and luminal surface have sugar transport properties which closely parallel those in the kidney and where the antiluminal membrane also has similarities to the exterior surface of the red cell, namely (a) relative insensitivity to inhibition by phlorizin, (b) independence from Na^+ concentration and (c) inhibition by phloretin and cytochalasin B [50].

On the basis of the foregoing data it seems appropriate to ask whether the proposed structural (and functional) link between conformational states of the sugar transporter spanning the red cell plasma membrane (shown diagramatically in Fig. 4.10), has a counterpart in the G and G' transporters which exist in the polarized proximal tubular cell (see Figs. 4.7 and 4.13). If this were true in the kidney, then the two glucose transport systems at opposite faces of the tubular cell might somehow be linked into a functionally co-operative unit. For example, the sequential selectivity illustrated in Fig. 4.13 could be associated with separate metabolic fates for *d*-glucose molecules originating from blood and urine (see below). In other words, this type of transport specificity would permit functional compartmentalization within the renal cell.

These speculative remarks serve to introduce the topic of how specificity of transport processes is integrated into the global activity of the whole cell.

4.5 INTEGRATION OF TRANSPORT AND METABOLIC SUBSTRATE SPECIFICITIES

As discussed by other authors in this series there are many examples of highly specialized membrane receptors. In each instance recognition of some chemical signal is linked to a particular cell response. e.g. antibody — cell reaction → blastogenic response; hormone — receptor interaction → increased intracellular enzymatic activity mediated by adenylcyclase and cAMP with consequences in cellular metabolic regulation.

In the case of a transported molecular species, the usual fate is incorporation into a metabolic pathway. Berlin has studied the manner in which the specificity of substrate — transporter interaction compliments the specificity of substrate-enzyme interaction [51]. One of the examples chosen was the intestinal sugar transport system. In this case, sugars entering the intestinal cell become incorporated into the metabolic pool and are phosphorylated via hexokinase.

Several interesting observations can be made. First, the specificity characteristics of sugar uptake (i.e. the transport system) and its subsequent metabolism (via hexokinase) appear to be the inverse of one another. For example, 2-deoxy-*d*-glucose is not actively transported and yet this deoxy sugar has greater affinity for hexokinase than *d*-glucose itself. A more detailed analysis of the sugar substrates transported by the small intestine (as outlined in [51]) reveals that the most critical points for binding to the carrier are at carbon positions 1, 2, 5 and 6. Moreover a pyranose ring structure in the *d*-configuration is also essential. For the hexokinase interaction, the most significant binding sites are at carbons 1, 3, 4, 5 (and 6 to a lesser degree). The pyranose ring is not essential.

Thus carrier and enzyme have unique yet overlapping specificity requirements. An important generalization that emerges is that the carrier protein(s) recognizes molecules susceptible to attack by intracellular enzymes at the reactive site and protects this reactive site during transmembrane translocation. Accordingly, Berlin designates the specificity of transport proteins as 'reactive-site directed'. In the special case of the intertinal uptake of sugars, the carbon 6 position represents the reactive site where phosphorylation occurs and the OH group at this position is also a critical requirement for recognition by the transport system.

Extending this hypothesis to dog kidney, our data on the specificity characteristics of sugar transport (see above) suggest that it is the glucose molecules that enter the proximal tubular cell from the brush border (luminal) surface that are destined for phosphorylation by hexokinase.

4.6 SUGAR TRANSPORT IN TISSUE CULTURE

Sugar uptake has been characterized in a variety of mammalian cell culture lines. The data from chick embryo fibroblasts [52–56] demonstrate the presence of a single facilitated carrier mechanism shared by *d*-glucose, 2-deoxy-*d*-glucose, glucosamine, *d*-galactose and 3-0-methyl-*d*-glucose. The transport is inhibited by cytochalasin B. More recent evidence [57] indicates the existence of two rather than one distinct sugar transporter: a low affinity system ($K_m \simeq 3$ mM), shared by *d*-glucose, 2-deoxy-*d*-glucose, and 3-0-methyl-*d*-glucose; and a high affinity system (K_m 0.05–0.2 mM) used by *d*-glucose and possibly 2-deoxy-*d*-glucose. These two systems are found to be experimentally distinguishable by virtue of differing sensitivity to thiol reagents. The high affinity transport mechanism is completely inhibited whereas the low affinity system is not affected by N-ethylmaleimide.

These tissue culture systems provide fascinating experimental models with which to study cell regulation of membrane transport activity. For example the following manipulations have been carried out:

(1) oncogenic transformation by DNA or RNA viruses [52];

(2) exposure to fetal calf serum [53], and

(3) hexose starvation [54,57].

Enhanced sugar uptake has been measured in response to each of the above perturbations — the effect being the result of accelerated transmembrane transport. Although controversy does exist in the literature, the evidence seems strong that augmented transport activity is derived from an increase in the number of available transport sites (i.e. an increase in V_{max}) rather than changes in transporter–substrate affinity (i.e. K_m).

Serum stimulation of sugar uptake in fibroblasts is biphasic, with the second peak exhibiting post-transcriptional and post-translational regulation [53]. In contrast, the hexose starvation-induced increase in sugar transport is regulated at the transcriptional level and is inhibited by intracellular 2-deoxy-d-glucose-6-phosphate [54]. The implication is that this phosphorylated metabolite represses or derepresses transcriptional events which in turn alter the membrane density of available transport sites. In addition, the data of Christopher *et al.* [57], suggest that hexose starvation derepresses especially the high affinity, N-ethylmaleimide-sensitive sugar transport system.

Finally, passive infection of African green monkey kidney cells [58] with Simian virus 40 causes no differences in either α-aminoisobutyric acid or 2-deoxy-d-glucose transport. Thus lytic viral infection is not associated with transport stimulation as is observed with cell transformation. Therefore the transport stimulation observed in the latter circumstances must be a consequence of changes in host phenotype brought about by the transformed state rather than by the viral genome.

The conclusion we reach is that carrier *specificity* as well as *density* can be altered by a multitude of environmental influences acting at either face (interior or exterior) of the plasma membrane. Any changes in the delicate network of feedback controls operating between nucleus, cytoplasm and membrane may upset normal function and ultimately lead to pathological states such as neoplastic transformation [56].

4.7 INSULIN-SENSITIVE GLUCOSE
TRANSPORT MECHANISMS

d-Glucose enters muscle [59] and fat cells [60, 61] by facilitated stereospecific, saturable, carrier mechanisms which exhibit competitive inhibition, counter-transport, and are stimulated by insulin. The action of hormone in these tissues is mediated via specific binding to membrane receptors [62, 63]. However, despite much research effort, we still have only a rudimentary understanding of the molecular events associated with the insulin-induced stimulation of glucose transport.

Czech and co-workers [64] have employed preparations of isolated intact white and brown fat cells to investigate insulin-dependent transport. Based largely on kinetic studies involving the non-metabolizable sugar 3-0-methyl-*d*-glucose (which shares the *d*-glucose transporter in fat cells), Czech claims that insulin increases the V_{max} of the system without altering the K_m. These findings conflict with earlier work on adipose and muscle tissue which showed significant insulin-induced changes in transport affinity constants [60]. Making use of a new method to measure glucose transport in rat adipocytes, Taylor *et al.* [65] have demonstrated an inverse relation between intracellular 3', 5'-cyclic AMP (cAMP) levels and glucose transport. Kinetic studies reveal that as the cyclic AMP level increases the V_{max} of transport decreases. Conversely, insulin, which lowers intracellular cyclic AMP, increases the V_{max} thereby stimulating glucose uptake.

As already mentioned there is very little direct information as to the intramembrane events which occur secondary to hormone—receptor complex stimulation of transport. Czech's studies [66], however, do reveal that the insulin effector system is much more sensitive to thiol reagents than the transporter. Activation and deactivation of glucose transport appear to be mediated by two separate classes of sulphydryls: those close to the membrane surface are involved in activation by insulin whereas deactivation is associated with groups deeper in the membrane core.

Chang and Cuatrecasas [67] have shown that exposure of fat cells to low concentrations of ATP, but not other nucleoside triphosphates, inhibits insulin-stimulated glucose transport. The effect appears to be transport-specific since neither insulin binding nor its antilipolytic action are altered. This study suggests that phosphorylation of certain specific membrane const[i]tuents may be directly involved in insulin stimulation of carrier-mediated glucose transport. Since cAMP stimulates phosphorylation via activation of protein kinases [68] the results of Chang and Cuatrecasas are consistent with the finding that cAMP has an inhibitory effect on glucose transport in fat cells [66]. It then seems reasonable to postulate [66] that the ability of insulin to enhance glucose uptake is mediated via its lowering of intracellular cAMP levels.

The mechanism by which insulin stimulates glucose transport is a special example of a much more general membrane phenomena—the manner in which receptor—hormone complexes regulate membrane enzyme activities (e.g. adenyl cyclase) as well as transport proteins.

Using as a backdrop the fluid mosaic model of membrane structure [6] Cuatrecasas has advanced a two-step dissociable reaction sequence to account for hormone action [69]. First, specific polypeptide—receptor binding occurs. This induces conformational changes in the receptor so that special affinity is acquired for other membrane component(s). The model does not require receptors to be part of or contiguous to the effector molecules but instead leans heavily on the dynamics of protein interactions within the plane of the fluid bilayer.

Insight into intramembrane molecular rearrangements that might underlie certain biologic activities have recently been obtained using morphologic techniques. Freeze-fracture studies [70] show the presence of insulin—ferritin 'clusters' on fat and liver cell membranes. These clusters disperse upon exposure of the membrane to native insulin, suggesting that their formation might have physiologic significance. Formation of such receptor aggregates may provide a useful working model for explaining observed co-operative binding kinetics as well as hormone modulation of transport specificity [71].

4.8 EFFECT OF DRUGS ON TRANSPORT SPECIFICITY

Baker and Rogers [72] have studied the effects of phenothiazines on sugar transport in human erythrocytes. Increasing concentrations of chlorpromazine at first accelerate glucose exit and then cause progressive inhibition. Sorbose entry shows a complicated two-phased inhibitory behavior in the presence of this drug. The effect of chlorpromazine is that of a non-competitive inhibitor which means that it acts on the carrier without affecting hexose binding. However, when the drug concentration reaches 10^{-4} M the affinity of the transporter for glucose is also changed.

Other phenothiazine derivatives as well as imipramine and haloperidol exhibit similar actions on hexose transport.

The chlorpromazine–membrane interaction is consistent with hydrophobic bonding [73]. Therefore, the drug effect could be mediated directly via interactions localized to hydrophobic portions of the carrier or indirectly through adjacent phospholipids.

These rather subtle drug-induced membrane alterations represent yet another way in which membrane perturbations by exogenous agents can lead to changes in transport specificity.

4.9 CONCLUSION

In the light of the foregoing it seems hardly necessary to emphasize that experiments carried out on transport systems *in vivo* or *in vitro* reflect, primarily, the state of the membrane as determined by the environmental conditions at that particular moment in time. Interpretation of the data must be based on these conditions. Extrapolation to other experimental situations may be hazardous without confirmatory data.

Thus specificity characteristics for membrane transport systems are as unique as their constituent proteins and the local microenvironment. Therefore there is no virtue in trying to overgeneralize details from one tissue or species to another. Nevertheless characterization of the specificity of transport systems in intact tissue can be put to very practical use in later identification of component moieties during stepwise isolation and reconstitution procedures. In this chapter no attempt has been made to emphasize this type of application. Rather we have tried to explore the relevance of transport specificity determinants to the functioning of the intact cell.

Consideration has been given to the manner in which complementarity of substrate specificity requirements for carriers and substrate specificity for intracellular enzymes produces unique and selective metabolic reaction pathways. We have also discussed how complementarity of membrane transport carriers at opposing surfaces of polarized cells (e.g. kidney proximal tubule) leads to highly selective transepithelial reabsorptive pathways. In addition, we have given a brief description of the modulation of transport specificity by intracellular controls, both nucleic and cytoplasmic, as well as by polypeptide hormones and drugs.

When viewed from this perspective, the specificity of membrane transport systems is seen as a property which enables individual cells to achieve intracellular compartmentalization of molecular substrates. The development of such integrated macromolecular chemical networks thereby helps the organism to achieve increasing internal organization, and to decrease its entropic state at the expense of the environment.

At the macroscopic level, specificity and kinetics are seemingly easily dissociated one from the other. However, it is worth emphasizing that this separation may be entirely artificial. Thus expression of transport specificity may be inextricably bound up with kinetic considerations and the 'biology' of membrane macromolecular structures may only become understandable when we are able to operate at the microscopic level of molecular dynamics.

REFERENCES

1. Fischer, E. (1894), *Chem. Ber.,* **27**, 2985.
2. Pauling, L. and Delbrück, M. (1940), *Science,* **92**, 77.
3. Koshland, D.E., Jr. (1958), *Proc. natn. Acad. Sci.,* **44**, 98.
4. Jencks, W.P. (1975), In: *Adv. Enzymol. and Related Topics,* A. Meister, ed., J. Wiley and Sons, Toronto, **43**, 219.
5. Burgen, A.S.V. (1966), *J. Pharm. Pharmacol.,* **18**, 137.
6. Singer, S.J. and Nicolson, G.L. (1972), *Science,* **175**, 720.
7. Kornberg, R.D. and McConnell, H.M. (1971), *Proc. natn. Acad. Sci.,* **68**, 2564.
8. Linden, C.D., Wright, K.L., McConnell, H.M. and Fox, C.F. (1973), *Proc. natn. Acad. Sci.,* **70**, 2271.
9. Singer, S.J. (1974), *Ann. Rev. Biochem.,* **43**, 805.
10. Thomas, L. and Kinne, R. (1971), *Biochim. biophys. Acta,* **255**, 114.
11. Thomas, L. (1972), *FEBS Letters,* **25**, 245.
12. Silverman, M. (1973), *J. clin. Invest.,* **52**, 2486.
13. Mueller, P. (1975), In: *Energy Transducing Mechanisms* E. Racher, ed., Butterworth's, London, p. 75.
14. Krasne, S., Eisenmang, E. and Szabo, G. (1971), *Science,* **174**, 412.
15. Kytes, J. (1972), *J. biol. Chem.,* **247**, 7642.
16. Skou, J.C. (1964), *Progr. Biophys. mol. Biol.,* **14**, 133.
17. Kytes, J. (1975), *J. biol. Chem.,* **250**, 7443.
18. Kytes, J. (1974), *J. biol. Chem.,* **249**, 3652.
19. Ruoho, A. and Kytes, J. (1974), *Proc. natn. Acad. Sci.,* **71**, 2352.
20. Henderson, R. and Unwin, N. (1975), *Nature,* **257**, 28.
21. Oesterreit, D. and Stoeckenius, W. (1973), *Proc. natn. Acad. Sci.,* **70**, 2853.
22. Racker, E. and Hinkle, P.C. (1974), *J. mem. Biol.,* **17**, 181.
23. Silverman, M. and Black, J. (1975), *Biochim. Biophys. Acta,* **394**, 10.
24. Dodge, J.T., Mitchell, C. and Hanahan, D. (1963), *Arch. Biochem. Biophys.,* **111**, 119.

25. Seeman, P., Kwant, W.O. and Sauks, P. (1969), *Biochim. Biophys. Acta,* **183**, 499.
26. Lumsden, C.J. and Silverman, M. (1974), *Physics Today,* **27**, 9.
27. Lumsden, C.J., Silverman, M. and Trainor, L.E.H. (1974), *J. theor. Biol.,* **48**, 325.
28. Lumsden, C.J., Silverman, M. and Trainor, L.E.H. (to be published).
29. Elsas, L., Hillman, R., Patterson, J. and Rosenberg, L. (1970), *J. clin. Invest.,* **49**, 576.
30. Elsas, L., Busse, D. and Rosenberg, L. (1971), *Metabolism,* **20**, 968.
31. Chinard, F.P., Taylor, W.R., Nolan, F. and Enns, T. (1959), *Am. J. Physiol.,* **196**, 535.
32. Silverman, M., Aganon, M.A. and Chinard, F.P. (1970), *Am. J. Physiol.,* **218**, 735.
33. Kinne, R., Murer, H., Kinne-Saffran, E., Phees, M. and Sachs, G. (1975), *J. mem. Biol.,* **21**, 375.
34. Silverman, M. (1974), *Biochim. Biophys. Acta,* **332**, 248.
35. Silverman, M. (1974), *Biochim. Biophys. Acta,* **339**, 92.
36. Reeves, R.E. (1951), *Adv. Carbohydrate Chem.,* Academic Press, New York, p. 264.
37. Marsden, N.B. (1965), *N.Y. Acad. Sci.,* **125**, 428.
38. Prather, J.W. and Wright, E.M. (1970), *J. mem. Biol.,* **2**, 150.
39. Taylor, P.M. and Taylor, G.L. (1967), *J. biol. Chem.,* **242**, 14.
40. Silverman, M. (1976), *Biochim. Biophys. Acta* (in press).
41. Diedrich, D.F. (1966), *Arch. Biochem. Biophys.,* **117**, 248.
42. Burgen, A.S.V., Roberts, G.C.K. and Feeney, J. (1975), *Nature,* **253**, 753.
43. Murer, H. and Hopfer, U. (1974), *Proc. natn. Acad. Sci.,* **71**, 484.
44. Kolinska, J. (1970), *Biochim. Biophys. Acta,* **219**, 200.
45. Busse, D., Jahn, A. and Steinmaier, G. (1975), *Biochim. Biophys. Acta,* **401**, 231.
46. Silverman, M. and Huang, L. (1976), *Am. J. Physiol.,* **231**, (in press).
47. Barnett, J.E.G., Holman, G.D. and Chalkley, R.A. and Munday, K.A. (1975), *Biochem. J.,* **145**, 417.
48. Taylor, N.F. and Gagneja, G.L. (1975), *Can. J. Biochem.,* **53**, 1078.
49. Kahlenberg, A. and Dolansky, D. (1972), *Can. J. Biochem.,* **50**, 638.
50. Hopfer, U., Sigrist-Nelson, K. and Murer, H. (1975), *Ann. N.Y. Acad. Sci.,* **264**, 414.
51. Berlin, R.D. (1970), *Science,* **168**, 1539.
52. Kletzien, R.J. and Perdue, J.F. (1974), *J. biol. Chem.,* **249**, 3375.
53. Kletzien, R.J. and Perdue, J.F. (1974), *J. biol. Chem.,* **249**, 3383.
54. Kletzien, R.J. and Perdue, J.F. (1975), *J. biol. Chem.,* **250**, 593.
55. Weber, M.J. (1973), *J. biol. Chem.,* **248**, 2978.
56. Hatanaka, M. (1974), *Biochim. biophys. Acta,* **355**, 77.
57. Christopher, C.W., Kohlbacher, M.S. and Amos, H. (1976), *Biochem. J.,* **158**, 439.

58. Miller, M.S., Kwock, L. and Wallach, D.F.H. (1975), *Cancer Res.,* **35**, 1826.
59. Morgan, H.E., Regen, D.M. and Park, C.R. (1964), *J. biol. Chem.,* **239**, 369.
60. Crofford, O.B. and Renold, A.E. (1965), *J. biol. Chem.,* **240**, 3237.
61. Illiano, G. and Cuatrecasas, P. (1971), *J. biol. Chem.,* **246**, 2472.
62. Cuatrecasas, P. (1971), *Proc. natn. Acad. Sci., U.S.A.,* **68**, 1264.
63. Cuatrecasas, P. (1971), *J. biol. Chem.,* **246**, 7265.
64. Czech, M.P. (1976), *Molec. Cell. Biochem.,* **11**, 51.
65. Czech, M.P. (1976), *J. biol. Chem.,* **251**, 1164.
66. Taylor, W.M., Mak, M.L. and Halperin, M.L. (1976), *Proc. natn. Acad. Sci.,*
     U.S.A., **73**.
67. Chang, K.J. and Cuatrecasas, P. (1974), *J. biol. Chem.,* **249**, 3170.
68. Langan, T.A. (1973), *Adv. Cyclic Nucleotide Res.,* **3**, 99.
69. Cuatrecasas, P. (1974), *Ann. Rev. Biochemistry,* **43**, 169.
70. Orci, L., Rufener, C., Malaisse-Lagae, F., Blondel, B., Amherdt, M., Bataille, D.,
     Freychet, P. and Perrelet, A. (1975), *Israel J. Med. Sci.,* **11**, 639.
71. De Meyts, P. (1976), *J. Supramol. Struct.,* **4**, 241.
72. Baker, G.F. and Rogers, H.J. (1974), In: *Drugs and Transport Processes,*
     (B.A. Callingham, ed.), University Park Press, Baltimore, Md., p. 341.
73. Kwant, W.O. and Seeman, P. (1969), *Biochim. biophys. Acta,* **193**, 338.